Laszlo Roth is an instructor of package design at both the Parsons School of Design and the Fashion Institute of Technology. He was packaging director for Helena Rubinstein and the Ideal Toy Corporation. His work extends from packaging and product design to graphics and art direction for major U.S. corporations. He has illustrated over 20 books and is a regular contributor to NEW YORK magazine. Mr. Roth has received awards for his various design achievements and is a member of the Packaging Institute, U.S.A.

PACKAGE DESIGN

an introduction to the art of packaging

Laszlo Roth

A SPECTRUM BOOK

Prentice-Hall, Inc., Englewood Cliffs, New Jersey   07632

**This book is dedicated to my wife, Gabrielle,
for her invaluable assistance,
and to all my enthusiastic and talented
students at the Parsons School of Design
and the Fashion Institute of Technology.**

**Library of Congress Cataloging
in Publication Data**

Roth, László.
  Package design.

  (A Spectrum Book)
  Bibliography: p.
  Includes index.
  1. Packaging. I. Title.
TS195.R67   688.8   80-19253
ISBN 0-13-647842-5
ISBN 0-13-647834-4 (pbk.)

Editorial production/supervision by Heath Lynn Silberfeld
Interior design and chapter opening photographs by Graphikann
Manufacturing buyer: Cathie Lenard

10 9 8 7 6 5 4 3 2 1

Printed in the United States of America

PRENTICE-HALL INTERNATIONAL, INC., **London**
PRENTICE-HALL OF AUSTRALIA PTY. LIMITED, **Sydney**
PRENTICE-HALL OF CANADA, LTD., **Toronto**
PRENTICE-HALL OF INDIA PRIVATE LIMITED, **New Delhi**
PRENTICE-HALL OF JAPAN, INC., **Tokyo**
PRENTICE-HALL OF SOUTHEAST ASIA PTE. LTD., **Singapore**
WHITEHALL BOOKS LIMITED, **Wellington, New Zealand**

CONTENTS

FOREWORD

Life is a package. Of many forms... for many purposes. Plants, creatures, man. From seed, chrysalis, egg, or embryo through stages of use and performance to reuse, decay, even waste and recycling. Above all, it is incredibly complex.

Like life, packaging takes on many forms. Rarely does it exist only as a package. Its justification is the product within. The better it is done, the more it becomes a part of the product it serves. If we see it, we think we know it; seldom do we realize the many components of which it is made, the functions it serves, the arts and technologies from which it is drawn and sculptured. Above all, it is incredibly complex.

True greatness lies in utter simplicity. Whether the discipline is engineering, speech, corporate management, graphics, or writing. Few among us can take the chaos of knowledge and render the incredibly complex into the disarmingly simple.

Such a feat is this book. A primer—an act of vast simplification by one of those rare artists of technological skill, such that it serves to give breadth to the student, insight to the expert, and perspective to the specialist.

Above all this book is an exercise in learning. Facts are used to stimulate your creative skill and understanding to give you mastery.

Paul B. Reuman
Executive Director
The Packaging Institute, U.S.A.

PREFACE

The contents of this book offer a unique opportunity for the designer to gather practical knowledge of the most up-to-date developments in the fields of packaging and three-dimensional design.

Like the content of any course of basic instruction, it does not pretend to be more than an introduction to a field that is capable of almost limitless extension, personal variations of areas of creative activity.

The variety of materials available for the designer is immense: paper, cardboard, plastic, metal, glass, or wood. The designer can bring together the many techniques and methods of the diverse design field into a cohesive synthesis not only with the traditional skills of the hand, but with the ability to think in three-dimensional design terms.

This book is prepared for those interested in becoming professional designers of packaging, point-of-purchase and sales promotional items, and other three-dimensional products.

It is my sincere hope that this book will help students of graphic design as well as other creative individuals in search of understanding packaging and three-dimensional design.

acknowledgments

My sincere thanks to the following people who helped me in the course of preparing this book:

Jane Chope, Landor Associates

Jan B. Curtis, The Coca-Cola Company

Steve Doyal, Hallmark Cards

Harold Freeman, Packaging Engineer

Thomas Giaccone, Fashion Institute of Technology

Annette Green, The Fragrance Foundation

Fred Howard, Howard Displays

Charles Kirk, Photographer

Kenneth C. Leonard, General Foods Corporation

R. Bruce Holmgren, Editor, PACKAGE ENGINEERING

Sum Magdoff, Parsons School of Design

J. D. Mead, Procter and Gamble

W. F. Meyer, Container Corporation of America

Denise Ortell, Helena Rubinstein, Inc.

Richard Rabkin, Ideal Toy Corporation

Paul B. Reuman, The Packaging Institute, U.S.A.

G. M. Rekstad, Editor-Publisher, MODERN PACKAGING

William Ronin, Fashion Institute of Technology

John Russo, Parsons School of Design

Robert M. Schaeffer, National Paper Box Association

Thomas J. Serb, Fibre Box Association

Ethel Taub, Creative Displays

William J. Vitulli, The Great Atlantic and Pacific Tea Co., Inc.

Wallace Church Associates, Inc.

Leslie Weller, Elizabeth Arden, Inc.

I would like to thank my editor, Mary Kennan of Prentice-Hall, Inc., for her generous help and cooperation.

Laszlo Roth
New York City

CHAPTER ONE

the packaging industry

the packaging industry

Introduction to the world of packaging. What the package designer does. A brief description of the packaging industry. Facts and figures about today's packaging. The role of packaging in our changing society. Visual marketing and selling of products. American buying patterns. Functions of the designer as a member of a marketing team.

How many design-oriented persons understand the role of the package designer? Around the product, the designer must create a package that is unique and aesthetically pleasing; it must sell the product, run on existing packaging machinery, and represent a great step forward for the designer as well as for the client. The designer, then, must serve and satisfy five masters: the product makers, the package makers, the package regulators, the package sellers, and finally, the consumer. Only a designer with a broad background in design, marketing, and technical know-how can attempt this.

Ours is a technological culture with great emphasis on consumerism. This implies more than merely selling a product. It implies, as well, creating the need for the product, creating

1 Supermarkets promote products and entice customers with point-of-purchase displays.

1

the product itself, and packaging the product for mass distribution. With more than 1 million persons involved, the packaging industry employs more than any other single industry in the United States. More than 300,000 companies in 200 manufacturing industries perform packaging functions; these companies, in turn, need more than 5,000 supplier firms to provide packaging materials and equipment. Recent economic analyses by the Packaging Institute USA, the country's major professional packaging organization, indicate that the value of packaging in this country is over $43 billion (1978). Seventy-five percent of all goods purchased by consumers in the United States are distributed in packages.

Even with this widespread use,

2

most packaging represents only 5 percent of the value of all finished goods in the country. Packaging is definitely NOT the villain of our constantly escalating costs. In fact, Packaging Institute USA data prove that the costs of the finished goods would be probably five times as high were it not for the protective and distributive functions of packaging.

Packaging touches all phases of our existence. The manufacture, use, reuse, and disposal of packaging is an important factor of our daily life. Other seemingly peripheral industries, such as advertising, marketing, point-of-purchase (display), and promotional materials are dependent upon packaging. Point-of-purchase materials, in particular, are specifically created to engage the consumer at the point of

sale. The point-of-purchase display is one of the most effective forms of total visual marketing and selling.

The biggest effects on packaging have been caused by the changed demographics of age and longevity, income, the working woman, new life styles, increased leisure time, and the changes of our population's living habits from urban to

3

suburban|Our mobile generations pursue the great American heritage: the use of the automobile for travel and shopping. It was the widespread use of the automobile that established the shopping center with its multiple stores and services, now a core for American buying patterns. The ever-expanding supermarkets and retailing chains across the country have demanded a new look at the methodology of selling, packaging, and new product introduction to the consumer. Because each of these stores offers "more" and "better" for the consumer's dollar, truly effective, smart marketing strategies and planning are required. In the supermarkets, out of 10,000 items available, about 20 percent are sold at cost or less to bring shoppers to the store, where they will be enticed to buy up to 50 percent on impulse. This concept of the "loss leader" is one outstanding example of creative marketing.

The competent package designer knows that creative marketing is aware of coming trends, and responds accordingly. Knowledge of the buying habits and attitudes of the changing consumer can be of great help in planning the graphic and structural design of the package or the point-of-purchase display. Design today is a precise science, and the designer is an integral part of our contemporary marketing, helping the producers to create products, selling devices, and packages.

A well-planned packaging program will require the teamwork of many experts—graphic and product designers, people in the areas of marketing, sales promotion, manufacturing and distribution. As a member of a team, the designer must integrate in a design the various needs of all these experts. The package design must be effectively tied in with newspaper, magazine, and television advertising.

Too many designers in packaging and three-dimensional design have not absorbed the needs and wants of a changed society—the shifting demographics, the new technologies, the role of the computer in design, government intercession as represented by the FDA and the FTC, new awareness of ecology, and the problems of recycling packaging materials and consumerism. The designer will have to give serious attention to these new developments, many of which can be converted to design and marketing advantages.

suggested reading

PACKAGE ENGINEERING, 1978, an influential trade publication. Packaging Institute USA, annual reports of this important trade association.

CHAPTER TWO

aesthetics and tools

aesthetics and tools

aesthetics and tools

You as a designer. How the designer works.
Where ideas come from. From idea to the visual
image and product. The process of designing.
The methodology of design research and design
development. How to communicate a design
concept to a client. Training to be a designer.
Importance of the arts of drawing, rendering,
photography, and lettering. Tools and materials.
The study of typography. Description of the
basic typefaces, composition, and copyfitting. A
word about lettering. About color. Photography
as a design tool. A short history of photography.
Job possibilities in photography. How the
professional photographer works. About
equipment. How to design a package with a
camera. Your first photo-design assignment.
About retouching. Illustration options. Reflective
art and transparency art. How to prepare a slick
package comp. The three phases of a design
project.

the making of the designer

Designing is an art, and the good designer is an artist. The actual work of the designer begins with the IDEA. Where does the idea come from? Who thinks of it? How is the idea developed into a visual image? How is the visual image translated from a basic sketch to a three-dimensional object or concept? These questions and their answers deserve consideration and careful study, because they form the basis of creativity, the all important asset of the designer.

Leonardo da Vinci was one of the most inspired creative thinkers in the history of the human race. Art received only part of his attention. Leonardo was the first modern man. He observed life, testing his theories. He was a scientist, painter, inventor. It was the general opinion among his contemporaries that whenever Leonardo undertook a creative task, he would produce something unusual, something wonderful to behold. And he did. The notebooks of Leonardo constitute a repository of 5,000 pages of extraordinary artistic and scientific research and inquiry. Here we have sketches of scientific instruments, hydraulic engines, sections of bones, muscles, leaves, trees, and cloud formations. The more he studied, the deeper his wisdom, the sharper his skills became. Each new undertaking implied a new and

4 Drawing by Leonardo DaVinci. Reproduction, from the author's collection.

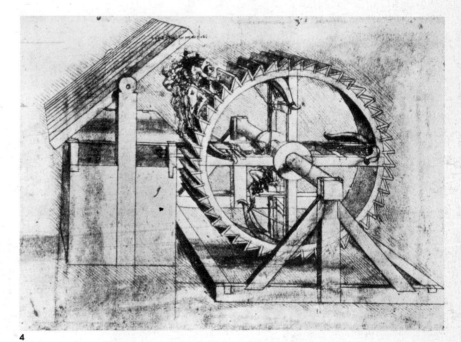

4

unique design. While he did not work by a formula, his creativity is based on careful, systematic observations and studies.

legwork

Designing for commerce and industry is a form of creating and selling products. In order to sell, the designer must know the product intimately, in terms of what the consumer expects from it. He or she must be thoroughly familiar with the needs, preferences, tastes, purchasing power, and buying habits of the consumer. The designer also must be aware of the client's needs and problems: the competition, marketing problems, and the client's budget must all be considered in planning the particular design project.

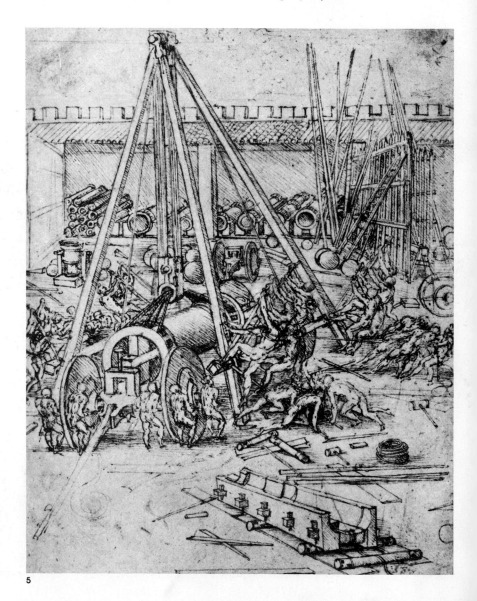

5

All this calls for extensive preliminary work before the actual project can begin. In most cases, marketing research and analysis are available for the designer. But this is not enough. The designer must study the existing market of the particular product by visiting retail stores and interviewing managers, sales personnel, and, if possible, consumers. The designer must then evaluate this information, comparing and analyzing views and opinions. Remember: a pretty package in the store is never ALONE. The package has a company of OTHER packages, competitive packages. The designer should see how they compare with his or her company's packages. Homework must be done carefully, methodically. The designer should read the trade papers and special books on the subject, and talk to packaging suppliers, who are always well informed of the latest techniques and materials. From this careful preparatory work, the designer may produce a visual concept, a promotional device, a package, or even a product. The core of these efforts is the IDEA, the visual aspect of it is the CONCEPT.

An idea can come from anywhere or anyone. However,

7

8

9

10

a specific idea for a specific purpose can only come from a trained, disciplined mind, accustomed to thinking realistically, and to rationalizing the visual aspects of the idea. The professional designer must be equally at home with AESTHETIC, TECHNICAL, and MARKETING aspects of the project. The professional designer is an artist who can draw, photograph, model, or build in order to present to the

client the TOTAL CONCEPT based on an idea.

Marketing is a science by which trained professionals develop products or the need for a product and plan sales strategy and sales campaigns. Packaging is also a marketing function, and the designer should be familiar with marketing procedures and techniques, either through reading or through courses on

the subject. Once a designer accepts a job, his or her first visit should be to his firm's or his client firm's marketing department to meet the group that will help him in the process of developing ideas.

Obviously then, the chief qualification of the designer is the ability to use skills to communicate the concept to the client. In addition to these physical skills, the designer must

11

12

7, 8, 9, 10, 11, 12 Handlettered labels. Courtesy of Push Pin Studios, Inc. (Seymour Chwast, designer).

have some knowledge of the history of art, especially the history of styles. The designer should be a reader, observer, and critic of the current scene—the theater, the cinema, the world of music. The designer should visit galleries, museums, private collections, and be exposed to all kinds of creative sources. These are where the ideas come from.

The professional designer should be able to draw. One of the best reasons for this is that visual concepts must be understood by people who are not visually oriented, but whose judgment and approval are necessary. Concepts are worked out in the final form by a professional illustrator or photographer. If the designer does not understand the technique required to produce the final art, this will create innumerable costly, time-consuming delays in a field where time and costs are the most important factors.

Sound training in the art of drawing is the first step to becoming a successful designer. There is a saying in the design field that one can make a designer out of an illustrator, but it is difficult to make an illustrator out of a designer. Some of our finest contemporary graphic

13 Artists-designers draw what they know—not always what they see. Courtesy of Helena Rubinstein, Inc.

14, 15, 16, 17 Different approaches to line and mass.

designers (Push Pin Studio Group, for example) are illustrators as well as superbly gifted designers. In recent years, all major art schools have been adding more hours of drawing to the curricula of their graphic arts students. The future designer should take as many drawing courses as possible.

Here is a suggested list of basic courses the designer should take:

1. A "heavy" drawing course (life drawing), at least 3 hours a week. Keep a notebook (sketchbook) and take it with you wherever you go. Draw whenever you can, everything and anything. Never tear out pages of "bad work"; keep it all together to see how you are progressing. Use any medium you prefer—pencils, markers, fountain pens, and so forth. Remember Leonardo's notebooks. Your notebooks are your own personal records of observations and studies.

13

14

2. Learn to work in various media—tempera, oil, pastels, acrylics, watercolors, and above all, markers. Take a course on rendering with markers AFTER you have learned how to draw. Marker rendering requires special techniques which only a seasoned draftsman will master.

3. Photography is absolutely necessary for the designer. Take a good technical course and buy a fine 35 mm camera with a projector. This basic equipment will be a tremendous asset and will help you in your work.

4. Lettering is definitely not a lost art. It is very much alive and it can get you great jobs. Brush up on your lettering and learn comp lettering. Don't always depend on ready-made type. Be original, be creative.

5. Cultivate the beautiful and develop your taste by learning about great art. Study the fundamentals of art history. Have an open mind, be interested in ideas, and above all, respect

deeply the life of your mind, your source of creativity.

There are designers, imitative and extremely verbal, who are in business more to make money than to improve design and taste. Other designers have the attitude that design is their personal expression of art (usually at the client's expense). Then there is the "yessir," "we can do it," "no problem," type who will do exactly what the client's whim dictates. (The client, of course, will get the same old cliché solution.)

The well-trained, intelligent designer is a professional who solves problems and does not create others. Packaging and design used to be in the hands of printers, suppliers, and brokers. Today's management is learning for itself what design is, and today's designers are better-trained professionals, concerned with design,

marketing, and sales. For the client, the designer can both establish graphic standards and be aware of cost and production problems. These are the criteria of the designer. It would be a crucial mistake to ignore these facts.

The dream of every creative artist-designer is to develop a "style" that is recognized, appreciated, admired, and imitated by all. The secret of developing a style and becoming perfect in all media is PRACTICE. Knowledge is the foundation of drawing. The artist draws what he KNOWS—not always (as is commonly suggested) what he SEES.

The artist is trained in HOW to see with the mind; the eyes are only his tools. While the painter usually works from life models, the designer must render quickly and simply without them. This fact increases the

15

16

17

18

18 Starting point: a place to draw, some basic tools, and an idea. A workable arrangement should be conducive to efficient work methods—and the ready development of new ideas.

19, 20, 21 Marker sketches by Cristina Marquez.

importance of KNOWING—knowing the essential structure of objects, landscape, and the human or animal figure. The artist must develop a perception that recognizes shapes and masses, not only lines, and he or she must learn the art of simplifying—the ability to distinguish between what is important and what is incidental and therefore irrelevant.

tools

The basic tools for the designer are simple. At the drawing board, all that are needed are some soft pencils (HB, 2B, 3B, and 4B) for quick thumbnail sketches (some designers prefer thin black marker pens); a soft eraser; colored pencils and assorted colored markers; designer colors (tempera); waterproof India ink (black); ruling pens (or rapidographs); a

19

20

21

compass; X-ACTO knife (no. 11); rubber cement and white glue; a steel ruler for cutting and scoring; and the extremely useful T-square and triangle. For visualizing, excellent papers are available in pads of different sizes. These layout or visualizing papers are suitable for pencil, inks, and markers.

One of the most important recent developments in graphic media is the marker pen. This simple tool has created a veritable revolution in the graphic arts. In the 1940s, the refillable felt point pen was developed. In the 1950s, the first color markers appeared, and became an almost instant success with artists and designers. Marker colors are formulated with dyes, not pigments. They come under the brand names of different manufacturers, some in more than 200 colors. Some makes are coordinated with Pantone® papers and printers' inks. Others are water-, thinner-, or acrylic-based. An infinite variety of markers is available, with tips ranging from extremely fine to a wide chisel point. Markers are versatile. The trained designer can achieve effects remarkably similar to pen-and-ink, wash, watercolor, pastels, and even oils. It is no wonder that markers have become the favorite tools of the designer. Rendering with markers is now a separate study in art schools, taught by experts in this medium. (Some marker techniques are shown in Figure 19, 20, and 21.)

All these wonderful media and tools will not make a designer or artist out of someone who will not follow the proper system of learning by doing. Art education is not all preparation. The only form of practical art education is experience—practice, sketching, drawing, learning, and exposure to all forms of arts and ideas. All the preparation in the world will only teach the designer HOW TO PREPARE.

the five don'ts:

1. DON'T work with short fumbling lines. Learn to use light or bold continuous line. Use soft pencils or markers, not pen-and-ink.

2. DON'T try to show too much. Close-ups of heads and shoulders are preferable to full figures. Suggested backgrounds are better than confusing details.

3. DON'T try too much perspective, except in its simple form.

4. DON'T get carried away with the drawing. Draw only to show the idea.

5. DON'T copy. There is no surer way to shut off all individual creativity and learning. You may, however, use and adapt pictorial reference material (swipes). In fact, start building your own picture files, which can save you a lot of valuable time in your design research. The "swipes" should inspire you, guide you, and help you to develop your own ideas through pictorial reference.

suggested reading

Donahue, Bud, THE LANGUAGE OF LAYOUT. Englewood Cliffs, N.J.: Prentice-Hall, Inc., 1978.

Gist, Ronald R., MARKETING AND SOCIETY. New York: Holt, Rinehart, Winston, 1974.

Kleppner, D., ADVERTISING PROCEDURES, 6th ed. Englewood Cliffs, N.J.: Prentice-Hall, Inc., 1979.

Nicolaides, Kimon, THE NATURAL WAY TO DRAW: A WORKING PLAN FOR ART STUDY. Boston: Houghton Mifflin Co., 1975.

Rosenberg, L., MARKETING. Englewood Cliffs, N.J.: Prentice-Hall, Inc., 1977.

Schiffman, Leon G., and others, CONSUMER BEHAVIOR. Englewood Cliffs, N.J.: Prentice-Hall, Inc., 1978.

Simmons, Seymour, and Marc S. Winer, DRAWING, THE CREATIVE PROCESS. Englewood Cliffs, N.J.: Prentice-Hall, Inc., 1977.

Troise, Emil, and Otis Port, PAINTING WITH MARKERS. New York: Watson-Guptill, 1972.

project

There are several ways of drawing a single object or a figure. The two better-known methods are the line and the mass. Take a single object or a figure and try to render it in line and in mass. The figure should be in action. Make at least five variations in both techniques. Draw from life, DO NOT COPY. Use pen or pencil for the line, chalk or brush for the mass.

22, 23 A single object or figure may be rendered using either one of the two better-known methods: the line and the mass.

22

23

typography

24 The essence of typography—a handlettered page by John Russo. Used with permission of the artist.

25 Condensed outline of root forms and development of the Roman alphabet, based on a single character. Thomas Blane Stanley, THE TECHNIQUE OF ADVERTISING PRODUCTION, 2nd ed., © 1954, p. 149. Reprinted by permission of Prentice-Hall, Inc., Englewood Cliffs, New Jersey.

Webster defines typography as "the style, arrangement, or appearance of typeset matter." Essentially, it deals with the ease and legibility with which we recognize letters and numbers. It deals with the right selection of specific typefaces for specific purposes; and it deals with aesthetics. Since typography appeals to the eye, the universal principles of design apply also to typography.

From a historical standpoint, typography is the study of not only the printed word but the written word as well. The earliest ancestors of our so-called "roman" type are the perfect, stone-cut capitals of Trajan's Column (114AD.) and the manuscript hands of Italy and Northern Europe (fourth to eighth centuries), which the first typographers purposely imitated and adopted. Through the centuries, variations of the

24

uncial and half uncial, the Caroline Minuscule (ninth and tenth centuries), and the Humanistic Book Hand (sixteenth century) gave us the forms on which all our types are based.

It is impossible to choose and specify type intelligently, unless we know many faces at sight. In order to accomplish that, we must study their present design in the light of the history and influences that gave them their shape and character. Following this approach, we can classify type into five groups: Text or Black letter, Roman, Italic, Cursive and Script, and Block Letters.

text or black letter. Text is the type used by Gutenberg and his contemporaries, developed from the fifteenth-century manuscript hands. Text is sometimes called "Old English" because it was used by Caxton, the first English printer.

roman. Roman type is the second group, also derived from the manuscript hands of the fifteenth century. In fact, it is to two German typographers, John and Wendelin of Spire, and to a Frenchman, Nicholas Jenson (all three working in Venice about 1470) that we owe this truly beautiful type.

italic. The type we know today as italic was first used about 1500 by the famous Italian printer Aldus Manutius. At first, italic was used with small roman capitals. The slanted or italic capital did not come into general use until about 1538, over twenty years after Aldus' death.

25

cursive and script. These styles are derived from informal running handwriting. Both cursive and script show a tendency toward linked letters and flowing strokes.

block types. This is a large group of typefaces, mostly without serifs and approximately of the same weight throughout. In the United States, such faces are called "Gothic." However, there is nothing Gothic about block characters, which are actually simplified Roman types of high visibility and legibility that were introduced in England about 1800. In our own time, a remarkably elegant type has emerged, based on block types: Futura. Like all block types, Futura comes in many varieties. More suitable for display than for body copy, Futura is spare, geometrical, and of contemporary design.

It is estimated that there are about 6,000 typefaces, enough to fill virtually any typesetting need. Typefaces come in sizes ranging from 6 to 72 points, with a complete font in each size. Variations are available in light, bold and extra bold, expanded, and condensed. Capital letters are called CAPS or UPPER CASE, the small letters LOWER CASE. The term "case" originated with early compositors, who kept the type in two different cases. In lower case letters, the upper stroke (as on the letter d) is called the ASCENDER; the downward stroke (as on the letter p) is the DESCENDER. The crossline at the end of the main stroke is called

the SERIF; typefaces without serifs are referred to as "sans serifs." The greater portion of the type is called the BODY.

The POINT and the PICA are two units of measure used in typesetting. The POINT, used to measure the height of a letter, measures .0138, or approximately 1/72, of an inch. The PICA is used for horizontal measurements of type. There are 12 points to one pica, or 6 picas to an inch.

composition

TYPOGRAPHY and TYPESETTING are terms which refer to the process of setting and arranging type. The earliest method of typesetting was done by hand. Machine metal typesetting was developed in the later part of the nineteenth century, when linotype and monotype were invented. Intertype and Ludlow were introduced in the early 1900s, and Photocomposition became available in the 1950s.

With the recent introduction of photocomposing machines, typesetting techniques are being revolutionized through the development of special computers for typesetting. Several different types of computers have been developed. These computers vary in cost, size, and capabilities, from those that

require a monitor to perform hyphenation manually to others that contain large memory storage systems that perform hyphenation electronically.

Proper type specifications should include all of the following detailed information: the name of the type; its weight (light, condensed, extended, etc.); whether roman or italic; whether all caps, small caps, or upper and lower case; the size of type (points) and the amount of leading (spacing) between lines; the width (in picas) of the desired line; and whether the copy should be justified (flush) or ragged left or right.

Copyfitting, or type calculation (often called "specking," which is an abbreviated form of "specifying") is a study in itself. The designer must have a comprehensive type specimen book and a collection of custom-styled (hand or photolettering) books. These books are available from typographers.

Good copy preparation and accurate "markup" insures accurate composition. Here are some of the rules:

All copy should be typed double-spaced, with wide margins, on only one side of an 8-1/2" × 11" sheet.

Copy markup should be done with colored pens or pencils to enable the typesetter to see instructions easily.

All instructions should be quite specific; if necessary, use standard proofreaders' marks.

The layout should accompany the copy, to give the typesetter a clear picture of the job.

It is most important to check and edit copy before it is set! Remember: the typesetter MUST set the copy as it is written. Spelling, punctuation, capitalization, and grammar should be carefully checked.

A great variety of "press type," under different brand names, is also available to the designer. Press type is display type, symbols, ornaments, and even simulated body type, printed on transparent sheets, that can be transferred to most smooth surfaces by burnishing. It comes most commonly in black and white, although some can be produced (at greater expense, of course) in any color. Adhesive-backed plastic lettering is also available in many sizes for permanent signage. This graphic product helps the designer quickly to create and prepare charts, slides, and package comprehensives.

Now, a bit of personal advice: Do not get addicted to press type. Some young designers actually design packages, logos, and headlines ONLY with available press type. A GOOD DESIGNER IS A FINE "COMP" AND HAND LETTERER. There is no substitute for original, creative work in typography and lettering.

suggested reading

Biggs, John R., BASIC TYPOGRAPHY. New York: Watson-Guptill, 1972.

Bruno, Michael H., POCKET PAL. New York: International Paper Company, 1974.

Gates, David, TYPE. New York: Watson-Guptill, 1973.

Goudy, W. Frederic, TYPOLOGIA. Los Angeles: University of California Press, 1940.

Leach, Mortimer, LETTERING FOR ADVERTISING. New York: Van Nostrand Reinhold Company, 1968.

Nesbit, Alexander, THE HISTORY AND TECHNIQUE OF LETTERING. New York: Dover Publications, 1957.

Schlemmer, Richard M., HANDBOOK OF ADVERTISING ART PRODUCTION. Englewood Cliffs, N.J.: Prentice-Hall, Inc., 1976.

Stanley, Thomas Blaine, THE TECHNIQUE OF ADVERTISING PRODUCTION. New York: Prentice-Hall, Inc., 1940.

project

Develop your own corporate symbol. Design your logo (no help from ready-made type, please). This special symbol must be suitable for stationery, business cards, products, packaging, and signage. It should be in color, but must look good in black-and-white.

As a starter, it would be a good idea to study and analyze some of the better-known corporate logos. For an exercise, start out with ONE word and see how it adapts to various uses (a business card, a vehicle, a billboard). Once you have solved this problem, proceed with YOUR corporate identity symbol.

color

Color in design is becoming more and more part of our daily life. For packages, printed matter, photography, movies, and television, color performs better than black-and-white. Color represents objects, scenes, and people with almost complete fidelity. In design, color can suggest abstract qualities. Color creates a pleasant (or sometimes shocking) first impression. Color has the psychological advantage of fixing visual impressions in memory and stimulating interest. Color gives prestige to the package, product, or advertising.

All these general reactions to color—attention, interest, and pleasure—constitute part of its sales value. We have been taught from childhood to make certain associations with certain colors. Some colors are likely to suggest particular ideas. The reds and oranges symbolize warmth, passion, war, danger; a host of ideas connected with action and life. Blood and fire are red; our source of life, the sun, appears as a circle of red, orange, or yellow. On the other hand, ice, snow, and water, as we usually see them, are bluish. Almost all colors of winter are tinged with blue. White is so closely associated with snow that its usual suggestion is cool.

The basic suggestions of "warmth" and "coolness" are widely used in design. A picture of an ice bucket looks cooler in cool whites or greenish blues than it would in oranges or reds. An electric heater would be rendered with reds, oranges, and yellows to suggest warmth and comfort. Other abstract impressions can be suggested by color. Purity can be conveyed by white, light blues, pale greens, or by tints associated in our minds with things that are pure, such as water, snow, and blue sky. Sky blue will frequently suggest serenity and peace. Joy or gaiety may be suggested by hot pinks, reds, and yellows or festive character often associated with parties, celebrations, and parades. Mystery seems to lurk in soft, dusky hues—deep blues, dark purples, and other colors definitely on the dark side. Deep reds, purples, and gold suggest richness and quality.

Of course, people's experiences differ, and we cannot be sure that a given color will suggest the same quality to all beholders. But we do know that the distinction between "warm" and "cool" colors is relatively constant.

Reaction to color is a pleasurable experience. In most people, the love of color is inherent. Colors that are both brilliant and brightly illuminated give us so much to see that we actually become "keyed-up," excited by the pure challenge to our faculties. We get the same pleasurable shock from colors that we get from music. Brilliant reds and oranges suggest loud lively music. Soft hues remind us of slow, peaceful passages.

Amazingly, much of life is rather colorless. It is true that bright colors are found in nature; but the flowers last just a few months, the rainbow rarely appears, the sunset lasts just half an hour. For every brilliant flower there are acres of somber brown earth. The brilliant blue sky can turn angry gray. Color is beautiful; it is loved and appreciated by everyone.

the science of color

Color used in printing is pigmental color—a paint. The difference between actual paint and printer's ink is the "vehicle" with which the coloring matter is mixed. The principal vehicle for oil paint is linseed oil, while that of the printer's ink is varnish.

Pigmental color differs from color produced by light in one important respect: the PRIMARIES, or the colors from which other colors can be produced. When a beam of sunlight is broken up by a prism into a band of color called the SPECTRUM, we see six major colors blending into one another: violet, blue, green, yellow, orange, and red. This classification is arbitrary. Sometimes a seventh color, indigo, is identified between the violet and blue. The trained eye can recognize other immediate colors such as blue-green, yellow-green, and so forth, at the points where one major color blends into its neighbor. All colors in the spectrum are based on three primary wavelengths: red, green, and blue-violet. From combinations of these, all other colors can be produced. In PIGMENTAL color, the three primary HUES which will give us all the others are crimson (magenta), yellow, and blue (cyan). When the three pigmental primary hues are mixed, an approximate black is the result.

HUE is color quality. When we distinguish red, blue, and yellow as colors without any qualifying terms (such as light, dark, weak, or strong), we are classifying colors by their hue. PRIMARY HUES are yellow, red (crimson, magenta), and blue (cyan). These hues cannot be produced by mixture. SECONDARY HUES are orange, green, and violet. These can be produced by mixture of equal parts of two primaries. INTERMEDIATE HUES are greenish-yellow and yellowish-green, bluish-green and greenish-blue, violet-blue and bluish-violet, violet-red and reddish-violet (purple), orange-red and

reddish-orange, yellowish-orange and orange-yellow. These hues are produced by a mixture of unequal parts of two primaries or of one primary and one secondary hue. In the intermediate hues, as the proportions of the mixture are varied, the color is naturally inclined toward the hue which has been most used. Additional subdivisions can be created, up to the point where even the trained eye finds it difficult to differentiate between hues.

VALUE in any color is its degree of lightness or darkness. Black is the lowest possible value, and white is the highest. There is no absolute black or white. Looking at color value, we find that hues themselves differ in value. For example, if we place a pure yellow beside a pure blue, the yellow appears lighter than the blue. The yellow is the more luminous of the two. (In fact, yellow is the most luminous color in the spectrum.) The value of all colors can be varied by adding white or black in different proportions. A hue lightened by the mixture of white is called a TINT; when it is darkened with black it becomes a SHADE.

CHROMA, or intensity, is the quality by which we distinguish strong from weak colors. Chroma is purity of color, that is, the color's freedom from neutral grays. A pure color

cannot be toned down with gray or black alone. To weaken a bright color, instead of adding only grays, the grays should be mixed with that color's complementary. For example, when a small amount of green is mixed with red, the effect is usually pleasant, and the toned-down red has more color quality than would be obtained with the black-white-gray mixture alone. As a rule, when a hue is mixed with a tint or shade of its complementary, the resulting tones are always subdued, pleasant, and close to the original color.

COLOR HARMONY is a matter of relationship of colors. Basically there are three methods of color harmony: harmony of various tones of the same hue, all in the same segment of the color wheel (monochromatic harmony); harmony of hues closely related but not identical; and harmony of complementary colors.

Attempts have been made to determine arithmetically how much space in a combination the different colors should cover. The principle of color balance is: the brighter and stronger a color, the smaller its area should be. This should be true for all color combinations. In recent years, with the development and use of luminous and fluorescent colors in printing as well as in the fine arts, overstatement has seemed to take the place of restraint. The scientist and the artist who are studying and working with color would agree that no mathematical formula can replace good taste and judgment.

The various color theories are not yet scientifically proven. Color theories are divided into three general groups that are based on different approaches and explain different effects. The theories of the physicists deal with the phenomena of light; those of the painters deal with the characteristics of pigments; and those of the psychologists deal with the "color sensations" that are transmitted by the eye to the brain. Color as used by the designer falls more into the second group, and the pigmental theories have received the most attention from them.

color systems and related equipment

The most significant color system development in recent years was the Pantone®* Matching System. The system has a range of 563 colors; each color number represents a formula that can be

*Trademark and servicemark of Pantone, Inc.

duplicated by any printer in the world. The Pantone® Color Specifier for designers is an essential tool. With this book, the designer can ensure that the printer has an accurate reference for any color in the system. Colors are shown on coated and uncoated stock and samples are supplied on perforated tabs which can be attached to artwork. Pantone® Color Papers (20" × 26"), available in 505 colors, are most suitable for packaging comprehensives. Pantone® Color Tint Overlays (20" × 26") are transparent, self-adhesive films that come in a range of 210 solid colors and 58 screen colors and can produce a variety of graphic effects.

Another remarkable color system device for preparing packaging and art comprehensives is the 3M Color-Key. Basically, this is a photographic process by which full color artwork can be produced from black-and-white art on transparent films for use as an overlay for the package or layout. There are more than 50 colors to select from, and other colors can be created with dyes or inks. The final color art can be either transparent or opaque. This system lends itself to various experimental graphic media, like posterization. The method is inexpensive, fast, and virtually indispensable to the

preparation of professional-looking comps.

To receive information about these graphic systems, contact your local art supply store or the manufacturer. Again, a word of advice: these remarkable new methods and equipment are made to help the designer to produce professional-looking comps in the shortest possible time. But the designer must have training and discipline to learn design fundamentals, before applying these shortcuts.

suggested reading

Itten, Johannes, THE ART OF COLOR. New York: Van Nostrand Reinhold Company, 1962.

Stanley, Thomas Blaine, THE TECHNIQUE OF ADVERTISING PRODUCTION. New York: Prentice-Hall, Inc., 1940.

project

Suppose you have to design a series of packages: a cereal box, a child's beach toy, a shampoo for blondes, a vitamin C package, and a country and western album cover. The basic color scheme of these packages is yellow, from cadmium to lemon yellow.

Some of the possible variations are: use of yellow as an "overall" effect (monochromatic rendering with contrasting logo and lettering); a black-and-white illustration or photo on yellow stock; a bright yellow handlettered logo as the main design element; use of and emphasis on such "symbols of yellow" as sun, sky, or plants.

photography

Photography probably began with the discovery, in the mid-1700s, that exposure to sunlight causes silver nitrate to turn dark. Some experiments in controlling this phenomenon were made in the early 1800s. As early as 1700, scientists had discovered that light, converging on a small hole in a darkened box (a CAMERA OBSCURA), carried to the inside wall of the box a soft image or reproduction of the view as seen through the hole. When simple lenses were substituted for the hole, the picture became clear and sharp.

It was only natural that the sciences of chemistry and optics should combine forces. In 1835, the Frenchman Daguerre discovered a means of developing photographic plates coated with silver iodide by means of mercury vapor. By 1839, he had also perfected a means to "fix" these images permanently on silver plates with a solution of sodium thiosulfate. Exposures required several minutes, but the quality of the pictures produced was exquisite.

Optics and photographic plates were much improved by the late nineteenth century, and new, more reliable photographic papers were manufactured. But popular photography really arrived with George Eastman's invention of flexible roll film in 1884 to

27 Photography of the past. LEFT TO RIGHT: daguerrotypes, tintypes, early photo, box camera.

27

28

replace glass plates, and in 1888, a simple box camera to use it.

Photography has come a long way since the days of Eastman's box camera. Color photography has really changed our way of seeing the world. It has brought about changes in design, fashion, advertising, entertainment, education, and the recording of current events.

In the United States, there are approximately 50 million cameras in use today; more than 2 billion photographs, in color and black-and-white, are made yearly. The variety of films is enormous, and cameras come in many makes and types, from simple amateur cameras to the most complex professional types. Tremendous quantities of silver are needed for the manufacturing of the film; thousands of accessories

must be produced to keep up with the demands of photography.

As a profession, photography offers a number of exciting aspects. Today's photographer is a trained creative artist, designer, illustrator, and technician. The photographer must have a great deal of versatility and adaptability for planning and coordinating pictures, shooting in the studio or on distant locations, working with models, props, and a variety of equipment. It is a demanding but rewarding career.

Since photography is an essential medium in design, it is taught now in all major art schools for the designer. Techniques of picture-taking and darkroom technology are both important, practical courses.

The designer must be acquainted with the work and equipment of the professional photographer in order to be able to work on various design assignments. For the designer, photography is essentially a form of illustration. It offers both great fidelity and a creative, experimental quality.

Photography can be a time saver. It is suitable for product shots as well as photographs of people, animals, landscapes. It can be done indoors and outdoors, under any light or weather conditions.

The photographer must have a good-sized studio, with darkroom and dressing rooms. The studio should be equipped with the proper equipment needed for various jobs—lights, studio cameras, backgrounds, and basic props.

Most photographers use a large, stable 8 × 10 or 11 × 14 studio camera when high accuracy, quality, and fidelity are required. For similar quality but less costly processing, the 4 × 5 or 5 × 7 is most suitable. For action, speed, and economy, twin-lens reflex (2 1/4 × 2 1/4) or single-lens reflex (35mm) cameras are very popular with photographers. Polaroid or other instant-print cameras may also be used for test shots before the final shot is taken. Certain regular cameras have special attachments to accommodate such a system.

project planning

You have to design a package for a doll. The package must have absolute fidelity to sell the product. Rather than using a drawing on the box, you choose photography: a child playing with the doll. You prepare a layout showing how you would like the final picture to look.

First, have a pre-production meeting with your photographer. Present your layout, describe your ideas about the set, lighting, mood, the appearance of the child model, and your budget. The photographer will help you to solve some of these problems. You will interview and choose the model sent to you from the model agency. (Another rule here: Do not use amateur models, no matter how cute they may be. Aside from possible legal problems, the professional model is trained to perform for you. Children are particularly sensitive under the conditions imposed by photographic modeling.) Now you select the model's wardrobe and supervise make-up and hair. (Pay special attention to the nails and missing teeth!) You are now an art director working together WITH the photographer and the stylist.

Once the model and photographer are ready, get a Polaroid test shot (if possible) of your suggested layout. After studying the shot, you may decide to do several variations. Don't forget: Film is cheap, time is money! Let the photographer shoot as many rolls as required, covering every possible situation from all angles. Sometimes an unexpected angle will give you the best picture. Since most photographers are working against time, you must make fast and correct decisions during shooting. The photographer will help you to make decisions.

The shooting should not take more than an hour or two, and you will probably get your "chromes" the next day. Next comes the difficult task of selecting the "right" chrome from more than fifty or sixty possibilities.

29

30

29, 30, 31, 32 The designer will choose the appropriate chrome for the package. Photo courtesy of Ideal Toy Corporation.

Once you have made your selection, examine it thoroughly under a magnifying glass. You may notice, for example, that the model's hair appears too light. This can be corrected by an expert chrome retoucher. Color correction on transparencies (chromes) is highly skilled work. Colors can be altered or changed from light to dark, colors can be added, patterns can be changed or removed. There

are some limitations, but a gifted retoucher can solve a lot of difficult problems.

If the transparency is a 35mm slide or a 2 1/4 × 2 1/4 chrome, a duplicate (dupe) must be made in a larger size—usually 4 × 5 or 8 × 10—to facilitate retouching. The retouching process basically consists of bleaching out areas to be lightened and adding dyes for deeper tones. Once the emulsion is removed from the chrome, it is impossible to replace it. Thus, most retouchers prefer to work on dupes.

It is also possible to make "assemblies," that is, combinations of several silhouetted chromes that are pieced together to form one large chrome. The preparation of these assemblies requires considerable technical skill, and is quite costly, but the result is usually excellent.

illustration options

The most important distinction between types of art is the difference between REFLECTIVE ART (paintings, illustrations, photographic prints) and TRANSPARENCY ART (slides or chromes made on color-positive films such as Kodachrome or Ektachrome). Reflective art more closely approximates the appearance of the printed page, and therefore affords more accurate reproduction. In transparencies, several hundred times more light passes through the highlights than through the shadows. Because of this, the transparency has considerably more detail and brilliance than reflective art.

Certain design problems cannot be solved or corrected on the chrome by the retoucher. Examples would be

31

32

special effects such as someone flying through the air, or the placement of a downtown view of New York City in the middle of a desert, or the addition of heads to bodies, or the simulation of stroboscopic motion photography. For such effects, it is necessary to have a very special print prepared, called the dye transfer. Dye transfers are the finest quality prints available for reproducing an image, either from a transparency or from reflective art. The highly controlled technique and dyes used in this process permit the closest match to the original color. The dye transfer is the only kind of print suitable for critical retouching. It takes several days to produce these prints, which are costly but remarkably accurate in detail and color. A skilled retoucher can achieve unbelievable effects on these prints.

More modest in price are the Ektacolor prints ("C" prints), which can be produced in hours. The quality varies. These are the most widely used color prints for packaging, layout, displays, and for many other visual aids.

Color photostats are extremely useful and economical for simple, basic presentations or comps, and a good Xerox-type duplicator can be used for many striking graphic effects. This is a useful piece of equipment for preparing layouts. The all-purpose photostat machine is also an excellent tool, if the designer knows how to use it properly. Besides reducing and enlarging, it can also be used to prepare and create original art by placing actual objects on the copy glass.

The decision whether to use photography or illustration is an often-debated issue among designers as well as their clients. It is true that photography suggests fidelity and truth, while illustration tends to be more individual in style, a direct statement of the illustrator. The fact is that each job presents its own specific problems. The designer should solve these problems by using the medium which is best suited to serve the causes of selling the product and of good taste and good design.

One of the best tools of the designer is the camera. A single-lens reflex 35mm camera is an invaluable aid in designing packages, sales promotional materials, and displays. A slide projector is a useful adjunct to the camera. With these handy tools, you are prepared to design any three-dimensional object.

Suppose you have an assignment to design an elegant cosmetic package for a light summer cologne. You think of textures, lovely views, a close-up of grass, flowers, abstract forms and shapes, moods. Your mind is full of beautiful ideas that would take weeks to develop into a format.

First, take your camera and try to find the places and things you want to shoot. Use a fast Ektachrome film, which can be processed in three to five hours if you live in a big city or near a good lab Now, take your slides and project them DIRECTLY onto the suggested blank carton, using the projector to reduce or enlarge the size of the image, to see how the shots work as a design. This is an excellent, time-saving method to come up with great visuals for a preliminary design presentation. With it, you can render your own photos for layouts, packaging, and even art for large displays, and your client will be impressed with your prolific pre-presentation.

Most design projects are presented and costed out in three phases. PRELIMINARY CONCEPTS are usually done with markers, in a loose but comprehensive style. This is the most creative and costly part of the project. THE COMPREHENSIVE (COMP) is a full-size structural and graphic presentation, flawless in appearance and suitable for photography for various uses (catalogs, sales aids, and other introductory materials). PRODUCTION MECHANICALS represent the final phase of the project, where all final art, photography, and illustration are prepared for production.

putting it together

In the preceding pages we have discussed in detail the aesthetic tools of the designer—the various techniques and equipment required to put together a slick, sharp, professional-looking comp. After much deliberation, handlettering was used for the logo. A "C" print, made from a chrome, was used for the illustration. The press type and the logo were produced in Color-Key. All this was applied to a paper carton with spray glue. To protect the comp, it was wrapped into shrink film (Saran) with the help of a hairdryer.

33

33, 34, 35, 36 Preliminary sketches are rendered in a loose but comprehensive style.

34

35

36

suggested reading

Asher, Harry, PHOTOGRAPHIC PRINCIPLES AND PRACTICES. Englewood Cliffs, N.J.: Prentice-Hall, Inc., 1974.

Hattersley, Ralph, PHOTOGRAPHIC LIGHTING. Englewood Cliffs, N.J.: Prentice-Hall, Inc., 1978.

Hattersley, Ralph, PHOTOGRAPHIC PRINTING. Englewood Cliffs, N.J.: Prentice-Hall, Inc., 1977.

Reedy, William A., IMPACT—PHOTOGRAPHY FOR ADVERTISING. Rochester, N.Y.: Eastman Kodak Co., 1973.

39

project

Design two packages with your camera:

1. A toy package for a pre-school toy. Use two small children for your models. (If you use amateur models, be sure that they play with the toy and do not look into the camera.)

2. A cosmetic package for a hair product. Can you change the color of the hair on your print? Try to use some transparent watercolors or dyes. For practice, try it first on a slick magazine illustration or color photo print mounted to a board. If you can handle an airbrush, this is a perfect way to utilize your skill. You can actually lighten dark areas with household bleach.

37 Comprehensives (comps) are full-scale, flawless, structural and graphic presentations, suitable for photography for various uses. Photo courtesy of Ideal Toy Corporation.

37

CHAPTER THREE

paper,
board,
and
structural
design

paper, board, and structural design

Introduction to structural design with paper and board. Some historical background about paper and papermaking. What you MUST know about paper and board, die-cutting, scoring, embossing, and adhesives. Folding cartons and set-up boxes. Charts and patterns of over 100 basic paper folding cartons, with detailed descriptions. The latest innovations in paperboard packaging. What you should know about corrugated containers. About the corrugated box industry. The various types of corrugated boards and containers. Another option for the designer: the Point-of-Purchase Display. Historical background. Some facts, past and present, about retailing. How to design a POP Display. POP categories. Your POP project.

Paper gets its name from PAPYRUS, a reed which the ancient Egyptians used for making writing material. Paper as we know it today was invented in China about 105 A.D. The first paper was made from the inner bark of the mulberry tree, pounded into sheets. Later, the inventors found that better paper could be made by pounding rags and hemp into pulp.

The Chinese art of papermaking spread to other parts of the world. A paper industry was established in Baghdad in 795 A.D. Following the Crusades and the Moorish conquest of Spain and North Africa, papermaking spread to all parts of the civilized world.

For centuries, paper was made by hand out of rag pulp. While this process was slow, the early demand for paper was very slight, although it began to increase after the invention of printing. In 1798, Nicolas Louis Robert, a Frenchman, invented a machine to make paper in continuous rolls, rather than in sheets, and by 1803 the English Fourdrinier family had taken over the financing and development of this machine. Even today, this papermaking machine is referred to as a "Fourdrinier."

In 1840, a German named Keller invented a process for grinding logs into fibrous pulp, called groundwood pulp. In 1867, an American, Tilghman, discovered how to separate the wood fibers by dissolving the wood in a solution of sulfurous acid. By 1882, the process of producing wood pulp had developed to a point quite similar to that used by modern paper mills.

There are three major processes for making pulp from wood: the sulfate, sulfite, and caustic soda processes. These are cooking processes whereby the pulp is washed, bleached, and cleaned of all foreign materials preparatory to feeding it into the various machines that will form the mass of fibers into a sheet. At this point, the pulp can also be colored with dyes. Glue or starch is added to give the paper a smooth surface for printing and writing. When a smooth, slick surface is desired, the paper sheet is covered with clay or other fine plastic material. Paper can be coated by impregnating it with waxes and plastics. Finally, the paper is passed through a set of iron rolls at the end of the paper machine to increase the smoothness and gloss of its surface. This process is known as CALENDERING.

38

38 PAPYRUS, a reed used by ancient Egyptians to make writing material, is the root of the English word PAPER.

characteristics of paper and board

All papers have a direction in which the fibers align. This is called the GRAIN. Paper or board will fold or score easily along the grain, but will crack when folded against the grain. Scoring with or against the grain makes no great difference, as long as the scoring is properly done with the proper tool. Paper or board torn WITH the grain produces a smooth edge; if torn against the grain, the edge will be ragged.

Since paperboard is made by laminating layers of papers together, any paper that is more than 0.012" thick (12 points) is called paperboard (or, commonly, cardboard). A popular weight for the average boxboard used in folding cartons is between 18 and 24 points. Illustration board is about 60 point. For displays or signs it is not unusual to use 80-100 point board. The fine Bristol board used for drawing, pen-and-ink, or watercolor comes in 2, 3, 4, 5 plies (12, 15, 18, 21 points). Board used for cartons is specified according to the size of the carton or, more often, according to the weight of the item that goes into it. The term CALIPER is used to denote thickness; it is expressed in units of thousandths of an inch (usually written decimally), which are referred to as caliper points.

papers

Different types and grades of papers are available for different uses. Practically every product area uses paper in some form. Some of the basic papers used in packaging are described below.

unbleached kraft is a coarse brown paper, the most economical and strongest packaging paper. Used for wrapping paper, paper bags, and general packing purposes, it can be laminated, coated, and impregnated with various protective materials.

glassine and greaseproof papers may be plain, printed, lacquered, waxed, corrugated, or laminated to other packaging materials. The main qualities of these papers are a water-vapor barrier, an odor and aroma barrier, and, of course, resistance to grease. About 85 percent of this paper is used for food packaging in various forms like pouches, bags, lined cartons, and envelopes.

parchment papers are manufactured by dipping sheets into a concentrated solution of sulphuric acid. The result is a tough, dense, translucent film that is sterile, free of fibers, strong when wet, and highly greaseproof. It makes an excellent liner or wrap for oily or wet items (butter, fish, vegetables).

tissue is used primarily as an inner wrap. Tissue may have a hard or soft surface. It can be waxed or impregnated with plastic resins for strength, and comes in translucent and opaque colors. It is widely used by florists and by the hosiery and food industries.

A number of papers are available for printing, labeling, and decorative packaging. Basically there are two types: the FLAT or DULL FINISH (coated and uncoated) and the GLOSSY FINISH. Smooth, high-gloss paper comes in brilliant white and colors; especially beautiful are the FLINT PAPERS, used for box coverings, labels, and gift wraps. They have good printability and some are scuff-resistant. The glossy-finish, SUPERCALENDERED whites (sulphites, clay, and chromecoats) emboss well and are excellent for high-quality printing.

METALLIC FINISH papers are called FOILS. These are metal foils laminated to papers or papers that are gravure-printed with metallic powders mixed with lacquers. They come in a wide range of colors and finishes and they can be embossed beautifully. They are used for specialty box coverings and overwraps.

A number of specialty papers are available textured with flock, glitter, or foam. IRIDESCENT and PEARLESCENT papers are used for box coverings and platforms (inserts) for gifts and luxury items.

boxboards

Boards account for approximately 60 percent of the total packaging market. Folding boxes, set-up boxes, and a wide variety of packaging devices are made of boxboards. With the exception of low-cost, plain chipboard, most boxboards are lined or laminated to another paper liner or liners. The inner board is usually 100 percent recycled fiber of low-grade papers. The outer or top liner varies according to need and quality. Some of the basic boards used in packaging are described below.

plain chipboard is the lowest-cost board, adaptable for special lining. It is not suitable for printing. The color range is from gray to tan.

white vat-lined chipboard has a white liner, adaptable for color printing. It is used for higher-grade set-up boxes with white liner.

bending chip is the lowest-cost boxboard for inexpensive folding cartons. It is usually light gray or tan, and can be printed in all colors.

bleached manila-lined bending chip is the same as bending chip, except the top white liner is of better quality.

white-lined 70 newsback is smooth board, much whiter than any of the above grades. The back is usually gray. It is excellent for folding cartons, displays, posters, and die-cut items.

clay-coated boxboard is a smooth, white board with an excellent printing surface. It is used for cartons, displays, and wherever high-grade, multicolor printing is needed.

solid manila board is available with white liner and Manila back for all types of cartons requiring durability and strength.

extra strength plain kraft-type boards are available with or without the white liner. They are used for hardware, automobile parts, housewares, toys, and other items requiring extra-strength packaging.

uncoated solid bleached sulphate is strong, white board, plastic-coated or waxed and water resistant. It is excellent under freezer conditions and is used widely for frozen food cartons.

clay-coated solid bleached sulphate has excellent printability, uniformity, and scoring and folding characteristics. Surface appearance (dull to high gloss) depends on coating process. It is most suitable for products such as cosmetics, drugs, textiles, and quality goods.

clay-coated natural kraft is a strong, moisture-resistant board with a white printing surface, used for heavy-duty packaging, bottles, beverage carriers, and similar items.

die-cutting

Every paper or board product, three-dimensional or flat, has a shape or form produced by die-cutting. Basically there are two methods of die-cutting. HOLLOW DIE-CUTTING is done with a die that is hollow like a cookie cutter. This process is used almost exclusively for labels and envelopes. STEEL RULE DIE-CUTTING is used where close register is required. It is a simple die. Steel rules are bent to the desired shape and inserted or wedged into a 3/4" piece of plywood. The multiple dies are locked up in a chase on a platen of the die-cutting press. Several sheets can be cut at one time. Flatbed cylinder presses can also be used for die-cutting.

One of the most important specifications in preparing a carton mechanical is the die sheet or "strike-out sheet." Never attempt to do a mechanical WITHOUT a die sheet from your printer, and never accept a size quoted over the telephone.

adhesives

Packaging materials need adhesives to join or fasten them together. An entire industry is based on the production of glues, cements, gums, and hot-melt adhesive materials. The creation and selection of the proper adhesive for the multitude of packaging materials and needs is a true science.

For example, the adhesive used for the popular self-adhesive or pressure-sensitive label is a synthetic latex formulation that can be used for removable or permanent labels. Certain clay-coated boards may require special resin-emulsion adhesives. Wet-waxed paper(board) requires more penetrating adhesive. Hot-melt adhesives are most suitable for adhering plastic films. Strict federal regulations govern the selection of the proper adhesive for food packaging. For industrial packaging when tapes are used, the tapes may be reinforced, pressure-sensitive, and in cloth, film, and foil.

Hand-made comprehensives should always be joined together with simple glue. Never use rubber cement for the glue flap. Rubber cement has a tendency to dry out, while white glue is a permanent adhesive. For mounting fabrics or thin paper, use a fine coat of rubber cement or spray glue.

scoring

A score is a crease made in the paper or board to facilitate folding. There are several ways of scoring. The most common method is to use a blunt-face (round-edge) scoring rule (die). The width of the rule varies with the thickness of the paper. When constructing a carton by hand, NEVER use a sharp blade for scoring. A cut will crush the fibers, which will weaken the paper. Use a blunt edge (ballpoint pen, paper clip, coin) against a steel ruler. Always bend AGAINST the score to produce an embossed, raised edge.

embossing

In embossing, the design or image appears in relief. This is achieved by pressing the paper to be embossed between a brass female die and a male bed or counter, both of which are mounted in register on a press. Embossing can be superimposed on printing or it can be done on blank paper for a sculptured effect (called "blind embossing"). Embossing is a costly process, generally used for prestigious packages, stationery, and promotional material.

40 In embossing, the design is created in relief by pressing the paper to be embossed between a brass female die and a male bed. Photo courtesy of Helena Rubinstein, Inc.

41, 42 Early paperboard packaging. Photos courtesy of Landor Associates, Industrial Design.

40

folding cartons

In 1879, a Brooklyn printer, Robert Gair, was inspecting a printed seed package which had been inadvertently cut on the press by an improperly positioned printing plate. The idea occurred to him to make up a die for cutting and scoring the paperboard in one impression, and to use a press to cut out carton blanks. Thus the folding carton we know today was born.

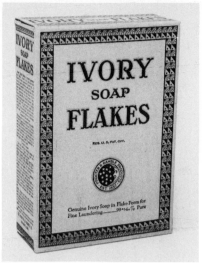

41

42

43 Early paperboard packaging. Photo courtesy of Landor Associates, Industrial Design.

44 Ladies' face powder compact, manufactured circa 1875, is a two-piece, drawer-style, extension-edge, rigid paper box.

45 Miniature hat box for doll's clothing, manufactured circa 1878, is a two-piece, shoulder-style, rigid paper box.

46 Garter box, manufactured circa 1870, is a hinged, shoulder-style, extension-edge box.

47 Soap sampler retail box, manufactured circa 1810. Both base and lid use tight-wrap-style construction.

Photos courtesy of National Paper Box Association.

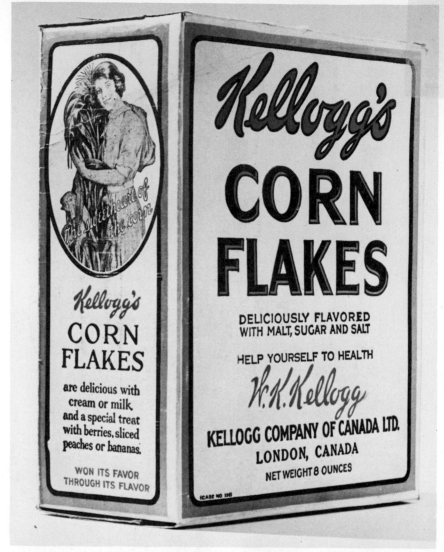

43

44

45

46

Today, folding paperboard cartons are a $2 billion industry, using about 2.5 million tons of board annually. About 530 carton manufacturers with 752 plants employ about 80,100 people, manufacturing over 250 million cartons yearly.

Folding cartons are precision-made, low-cost packages, supplied in knock-down form or "blanks." When erected, they become three-dimensional, rigid packages. They can be packed by high-speed, automatic equipment, or by semi-automatic or hand-operated equipment. These containers lend themselves to various marketing and retailing systems. They are shipped flat to the user and are not erected until needed.

Folding cartons are suitable for all known printing processes. The most used ones are offset lithography and gravure. Windows and die-cuts are common finishing operations. The high-speed equipment glues transparent films over die-cut windows in the blanks before they are folded. Special finishing and decorative treatments, such as embossing, varnishing, and texturing, can be applied before, during, and after printing. Application of wax, sealers, and laminations can be added to protect the carton's contents from moisture,

47

48

or to help prevent the contents from sticking to inner surfaces.

types of cartons

The choice of folding carton styles and constructions is dependent on the product to be packaged and the type of filling operations to be used. There are about five hundred styles and variations of carton constructions available to the structural designer; more are being added by highly skilled "paper engineers" every year.

There are two basic styles of folding cartons: the TRAY and the TUBE. Tray styles are cartons with solid bottoms hinged to the side and end walls. Sides and ends are connected by means of a flap, hook, locking tab, or lock. In addition to the basic tray, these cartons have a variety of cover and flap parts extending from the walls and sides of the tray. In another type of tray carton, two pieces, one slightly smaller than the other, form a base and cover of a two-piece "telescoping box." Typical tray packages are the ten-pack cigarette cartons, bakery trays, ice cream cartons, garment carriers, and a great variety of food cartons.

On the TUBE style, the body of the carton is a sheet of board folded over and glued against its edges to form a rectangular sleeve. It has openings on the top and the bottom which close with flaps, tucks, or locks. Tube-style cartons are used to package bottled products, cosmetics, pharmaceuticals, and a great variety of consumer goods. The tube-style carton gives the product fully enclosed protection. The addition of "windows" provides complete visibility of the

48 Folding carton "blanks" are made for a wide variety of three-dimensional packages.

49 Comprehensive for a folding carton. Design by Frank Hom.

49

50 Comprehensives for a line of men's fragrances.

51 Folding cartons. Courtesy of General Foods Consumer Center.

52 The consumer can see the entire contents in this folding carton. Photo courtesy of Ideal Toy Corporation.

53 A tray-style carton.

54 A tube-style carton.

50

51

52

53

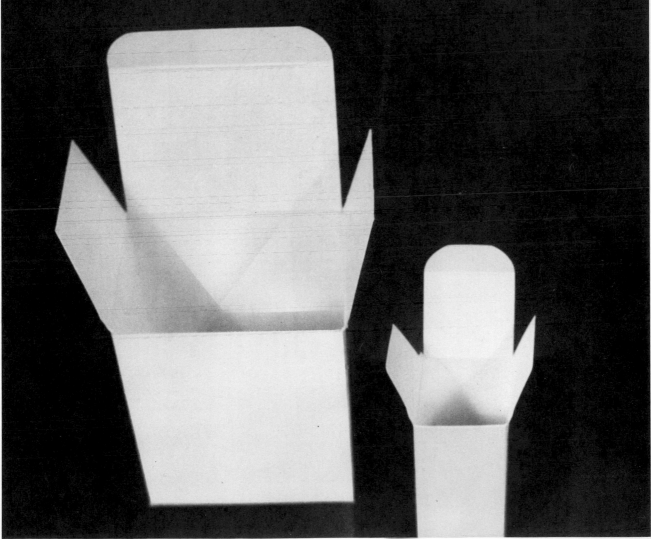

54

55

product. There is an infinite variety of unusual tube styles. Some are contoured, triangular, octagonal, or even rounded.

Among other innovative uses for folding cartons are overwraps for "multipacked" products such as individual portions of breakfast cereals, pet foods, and beverage carriers. The combination of shrink films with paperboards is

55 These Post® Grape Nuts® Flakes are one example of multipacked and overwrapped folding cartons. Photo courtesy of General Foods Consumer Center.

56 Typical paperboard packaging.

another method of packaging (see Chapter 4, "Plastics Primer"). Toys, housewares, and other contoured products are shrink-wrapped for display as well as for the protection of the product. Folding carton adaptations are the bag-in-the-box, boil-in-bag pouches, soups in pouches, and a variety of frozen foods in paper cartons that can be oven-heated and served in the package. Beverages such as milk and fruit juices are packaged in specially designed folding cartons that are lined with film and foil.

Innovation and the development of folding cartons have grown in the institutional market for food products. One system has a liquid-tight, leak-proof package that automatically folds to form a strong tray. This package can be frozen, stored, reheated in conventional ovens, used as a serving dish, and easily

disposed of. Similar packages are used in automatic vending machines that dispense both cold and hot foods.

The technology of the folding carton is an exciting study for the designer, the economist, and the sociologist. One of the greatest problems of history is how to feed the world. Packaging can provide part of the answer by designing containers to preserve and ship staple and perishable foods anywhere in the world—and even to outer space.

project

In order to understand the "anatomy" of the folding carton (tray or tube), it is helpful to construct several "blanks." With the help of the illustrations in the preceding section, construct several variations of your choice; use a strong Bristol board (2-3 plies).

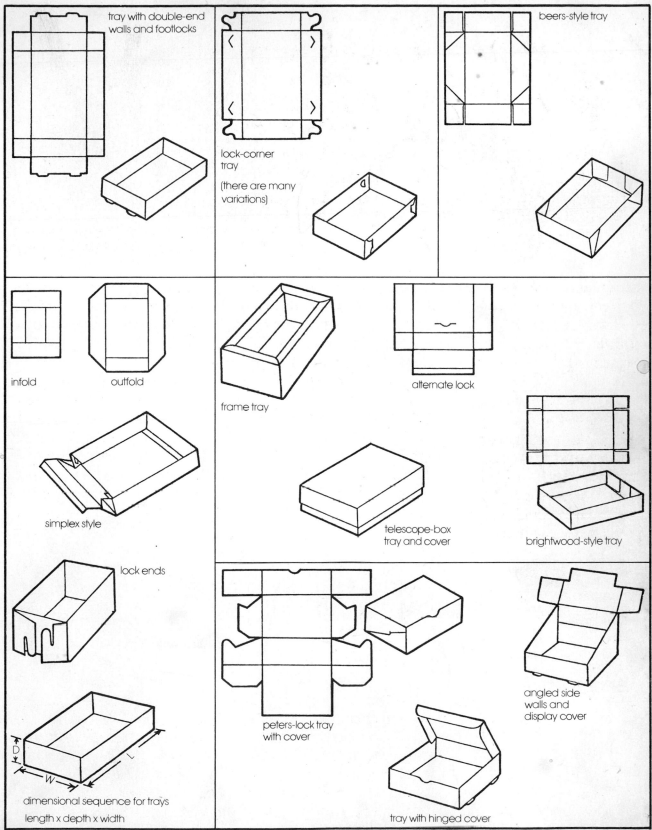

tray with double-end
walls and footlocks

lock-corner
tray

(there are many
variations)

beers-style tray

infold   outfold

frame tray

alternate lock

simplex style

telescope-box
tray and cover

brightwood-style tray

lock ends

peters-lock tray
with cover

angled side
walls and
display cover

dimensional sequence for trays
length x depth x width

tray with hinged cover

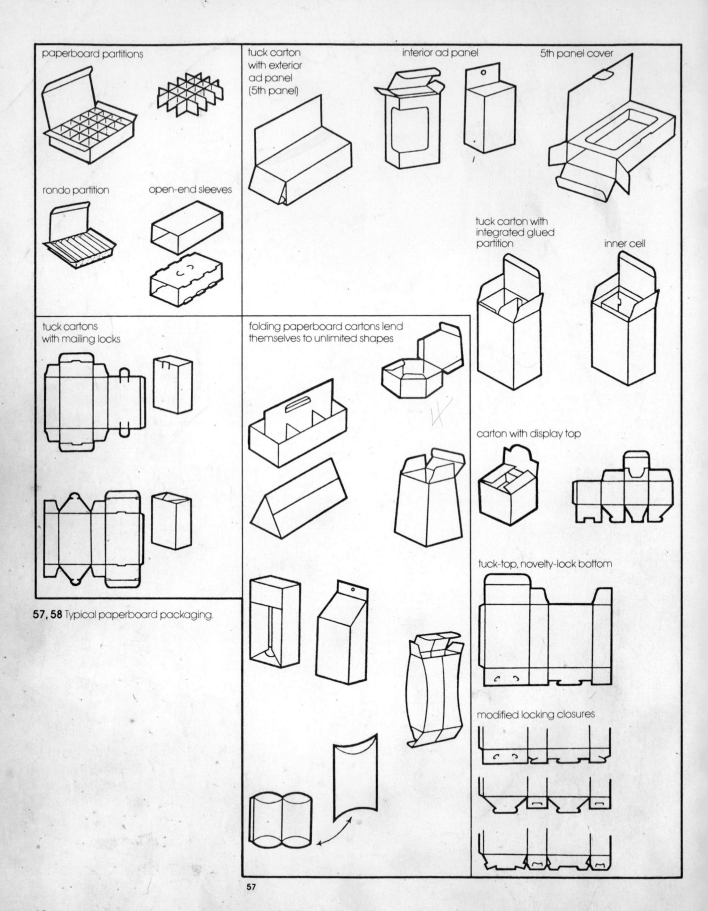

paperboard partitions

rondo partition

open-end sleeves

tuck carton with exterior ad panel (5th panel)

interior ad panel

5th panel cover

tuck carton with integrated glued partition

inner cell

tuck cartons with mailing locks

folding paperboard cartons lend themselves to unlimited shapes

carton with display top

tuck-top, novelty-lock bottom

modified locking closures

57, 58 Typical paperboard packaging.

57

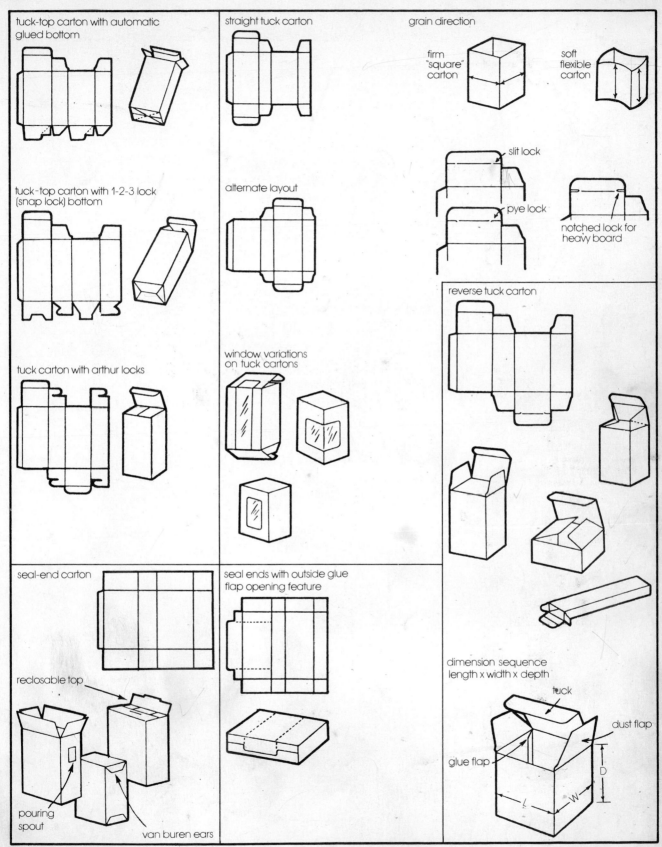

tuck-top carton with automatic glued bottom

straight tuck carton

grain direction

firm "square" carton

soft flexible carton

slit lock

pye lock

notched lock for heavy board

tuck-top carton with 1-2-3 lock (snap lock) bottom

alternate layout

reverse tuck carton

tuck carton with arthur locks

window variations on tuck cartons

seal-end carton

seal ends with outside glue flap opening feature

dimension sequence length x width x depth

reclosable top

tuck

dust flap

glue flap

D

L

W

pouring spout

van buren ears

set-up paper boxes

The history of the set-up paper box goes back probably to the invention of the paperboard. Since its first use, cardboard was used to make boxes covered with colorful paper wraps. These boxes were used by merchants to sell luxury products. Especially attractive boxes came from France at the end of the nineteenth century. Fragrances, ladies' hats, and fashion accessories were the typical merchandise. In Germany and England, toys and games were packed in cardboard boxes with colorful labels. In America, the first paperboard boxes were made in Boston in 1839 by Col. Andrew Dennison. His name is still synonymous with paper products.

Set-up boxes are rigid, permanent, three-dimensional containers, available in many attractive styles, shapes, and finishes. Two basic materials are used to construct rigid paper boxes: paperboard and covering materials. Many plastics can be combined with set-up boxes to make effective packages. Transparent plastic domes, windows, and thermoformed trays are among the options for use with set-up boxes.

There are many types and qualities of papers for box coverings. These range from inexpensive gift wraps to embossed foils and lacquer spray-coated sheets. If a more personal wrap is needed, a special design can be developed and custom-printed. Additional accessory materials such as seals, tags, ribbons, and ties are used to enhance rigid boxes. Set-up boxes are used for prestigious products—cosmetics, fashion accessories, jewelry, and cameras.

59 Set-up boxes are rigid, permanent, three-dimensional containers made from cardboard and covered with colorful papers or wraps.

corrugated containers

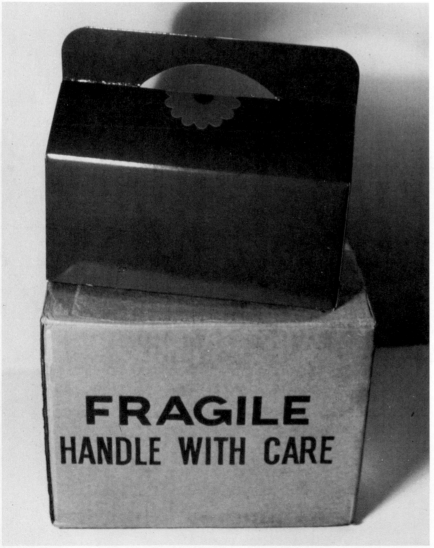

Corrugated board, like most packaging materials, has a long and colorful history. When you were a small child, this corrugated "box" was your favorite toy. As an adult, you packed your belongings in it when you moved to a new place. Your TV or other appliance arrived in an impressive corrugated box. It may be surprising to learn that this popular packaging medium, the workhorse of the industry, was originally part of an article of clothing. The famous gentleman's top hat of the nineteenth century was fashioned with a sweatband of fluted paper. An American inventor, Albert L. Johnes, patented this fluted medium for the protection of bottles in storage and shipment. In 1874 another American, Oliver Long, invented a process for sandwiching the flute between two paperboard sheets. This was the beginning of a new industry: corrugated containers.

Today, the corrugated container industry employs about 118,000 people in 1,427 plants and produces 200 million boxes yearly. It is a $10 billion industry. The corrugated container industry is now the largest industry of the paperboard packaging field.

The largest single market, representing more than a third of the industry's output, is for

60

foods. The list of users of corrugated is endless. It is used for shipment by every industry in the United States.

This extremely versatile packaging medium is basically simple in structure. It consists of a fluted sheet glued to one or more liners. A wide variety of combinations is possible, depending on the packaging requirements. The structural characteristics of the corrugated are governed by four variables: strength of the liners, strength of the corrugated medium, height and number of flutes per foot, and number of walls (single, double, triple, or more).

Four flute structures are available: A-FLUTE, which has greater capacity to absorb shock due to wider spacing of flutes; B-FLUTE, which has a greater number of flutes per foot to provide maximum crush resistance; C-FLUTE, which combines the properties of A- and B-flutes; and E-FLUTE, which is a very thin corrugated board. The E-flute corrugated box is perhaps the most popular type for large and sturdy display or decorative cartons.

Another important factor in corrugated box structure is interior protection. A wide range of corrugated partitions, liners, pads, and other devices, including plastics (molded

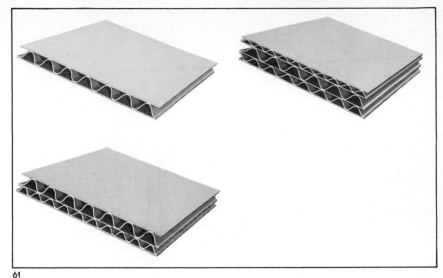

61

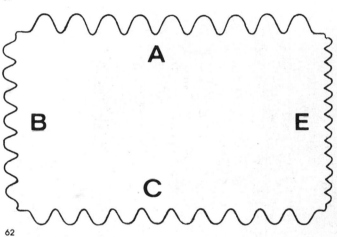

62

60 Corrugated containers are manufactured both plain and fancy.

61 Fluted sheets are combined according to packaging requirements. Photo courtesy of Fibre Box Association.

62 A "flute" gauge (actual size).

 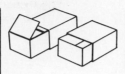 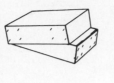 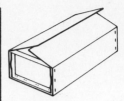 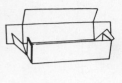

single- and double-lined slide boxes. **Single lined used as interior container and for parcel post, express, freight shipments. Double lined is three piece and provides double thickness all sides.**

triple- and double-slide boxes. **Triple slide (left) has two thicknesses of double-faced board all sides; double slide has two thicknesses on two sides. Collapsible; ideal for mail and express.**

telescope design box. **Extra thickness of corrugated board in side and end walls of this two-piece container affords exceptional stacking strength and overall protection to the contents of container.**

recessed-end box. **Three-piece box has a body sheet and two flanged end pieces. By varying body-sheet length, box size can change to fit many products of same girth but different lengths.**

five-panel folder. **Use for canes, umbrellas, similar long, slim items. Each end has a minimum of three thicknesses, providing strength where it is most needed. Container is shipped flat to user.**

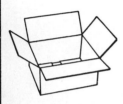    

regular slotted box (RSC). **Top and bottom flaps are equal length; folded inner flaps meet only if box is square. Securely sealed with adhesive, gummed tape, or metal stitches as desired.**

half-slotted box with half-slotted partial cover (pths). **Two-piece box, both sections slotted style. Double thickness of corrugated provides great resistance to bulging and buckling.**

full-telescope, half-slotted box (fths). **Full-depth cover, two-piece box. Both sections of slotted style. Full-cover top renders maximum product protection and superior stacking strength.**

center special slotted box (CSSC). **Construction gives double-flap thicknesses top and bottom. One or both side flaps are shorter than end flaps, so all flaps meet for double top and bottom.**

overlap slotted box (osc). **Efficient when products packed for shipment require sealing with metal staples, stitches, straps. Side flaps partly overlap for added rigidity at both top and bottom.**

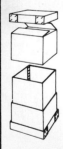  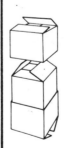  

double-thickness score-line box (Box With Cover). **Another design meeting requirements for double thickness score-line box under Railroad Shipping Rule 41. Box carries heavy loads despite rough handling.**

double-cover box **Popular with manufacturers of articles that cannot be readily packaged in standard containers. In large sizes, this box is often used as a unitized or palletized load.**

double-thickness score-line box (Conventional Slotted Style Box). **For high-density products (screws, nuts, washers) in weights to 300 pounds. Container is fast replacing wooden nail kegs.**

design box with cover. **Space-saving, stapled box with double end flaps and lid. Especially easy to pack, the design box is used for shipment of cut flowers, wreaths, and similar products.**

interlocking double-cover box (ic). **Flanges on covers interlock with flanges on tube. Three-piece box for items under Railroad Shipping Rule 41. Greatest use in packing heavy appliances.**

bellows-style box.
One-piece box closes by folding score lines at sides inward, pushing flaps down, sealing them with tape. Virtually leakproof; excellent for granular and powdered products.

full-flap slotted box (ffsc). Exceptionally strong. All flaps equal length, so when box length is exactly twice the width, end flaps meet to give both top and bottom triple thickness.

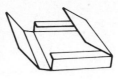

one-piece folder (1 pf).
Tucks of specified length. Container is easily stored, set up, packed, and closed. Used for parcel post and express shipments of books, apparel, and similar articles.

bliss box (no. 2).
Container offers considerable stacking strength. Used extensively for bulk packs of such products as meats, explosives, and other articles that have concentrated weight.

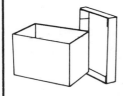

half-slotted box with cover (hsc). **Can be used as a combination shipper and shelf package and for various applications where box cover is required to be removed and replaced repeatedly.**

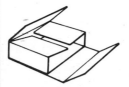

two-piece folder (2 pf).
Two scored sheets. Tucks of specified length. Two-piece folder is stronger than the one-piece design due to its thickness of board on top and bottom surfaces.

bliss box (no. 4). **When the body sheet of a No. 4 box is combined with the ends of a No. 2 box, double protection is obtained on all vertical edges. This container is called a 4-2 Bliss Box.**

double-wall slotted box.
This shipping container provides the extra strength and protection required for safe shipment of heavy products. It is made of double-wall board with various flutes.

three-piece folder (3 pf).
Three scored sheets. Tucks of specified length. Three-piece differs from two-piece by having two separate end tucks. Biggest demand is for long and flat products.

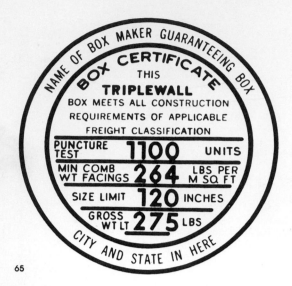

65 Weight, paper contents, puncture, and bursting test certificates must be displayed on all corrugated containers.

polystyrene), are utilized to provide inner reinforcing, cushioning, bracing, and shock absorption. MOST COMMONLY USED CLOSURES ARE STITCHING, STAPLING, GLUING, AND TAPING. (Figures 63 and 64 show some basic constructions.)

While some stock sizes are available, for special jobs a custom-made carton is usually required. Government and industry standards and regulations are designed to protect the corrugated user. These include laws pertaining to shipment by rail, freight, express, air freight, motor truck, and regular post. All, corrugated materials and cartons must be certified by the manufacturer. Weight, paper contents, puncture, and bursting test certificates must be displayed on all corrugated cartons. Some of the tests include drop tests, jolts, shocks, and vibrations to which a

product is subjected in normal handling. These tests are designed to select the right box for the product and the right "master carton" to ship the boxed product, without costly overpackaging or underpackaging.

Corrugated board can be impregnated and coated with various waxes and plastics. This method is one of the most significant trends in corrugated technology today. The moisture-protected coating permits the reuse of the carton for shipping products such as fruits and vegetables that were previously shipped in expensive wooden crates or barrels.

The most important advance in corrugated printing in recent years has been the use of flexographic printing, which utilizes quick-drying inks and high-speed presses. Another significant trend is toward greater use of colors on clay-coated white liner board. The use of pre-printed liners and full-color, lithographed, laminated labels (partial or complete) makes this workhorse of the industry an elaborately printed package.

suggested reading

Griffin, Roger C., Jr. and Stanley Sacharow, PRINCIPLES OF PACKAGE DEVELOPMENT. Westport, Ct.: The Avi Publishing Co., 1972.

Kelsey, J. Robert, PACKAGING IN TODAY'S SOCIETY. New York: St. Regis Paper Company, 1978.

MODERN PACKAGING ENCYCLOPEDIA.

THE PACKAGING INSTITUTE USA, additional information.

project

Test your structural skills. Construct a small carton from an A- or B-flute corrugated board. Devise some inserts to hold a small glass or glass bottle. Create a cushioning, bracing device to avoid breakage. Mail the package to yourself. If the glass arrives intact, you are an excellent structural designer. If it breaks, investigate the reasons and try again.

point-of-purchase displays

Point-of-purchase (POP) advertising is probably one of the oldest forms of communication. We have positive evidence that ancient Roman merchants often displayed and painted the symbols of their trade to attract their customers.

During the Middle Ages, these symbols were actual models of products, such as keys, tools, swords, utensils, and pottery.

With an illiterate consumer, obvious symbols had to be displayed. During the Renaissance, illustrated signs began to appear all over Europe—the first known collaboration between artist and tradesman. Some of these signs and symbols are still with us today. The three golden balls of the Medici (who were originally bankers and moneylenders) and the barber pole, which represented the

67

68   69

establishments of barber-surgeons in the days when the two professions were indistinguishable, are the most popular symbols, and ones which have survived for hundreds of years.

In the United States, the most typical nineteenth century point-of-purchase display was the Cigar Store Indian. This famous six-foot tall Indian was hand-carved from white pine, usually by carvers of ships' figureheads. The Indian was painted realistically and usually was placed in front of the cigar store. Collectors estimate that between 1850 and 1890 about 20,000 Indians were carved. Today the wooden Indian is a valuable collector's item.

In the early 1800s, merchants created their own sales aids with the help of a sign-painter. In the early pioneer days, the western saloon keeper and the general store merchant were constant sources of original signage, mostly based on contemporary, decorative typefaces. In New England, innkeepers produced the most original, often whimsical signs.

About the same time, brand names began to appear on barrels, crates, tins, and glass. Manufacturers began to supply store fixtures marked with their signs. It was considered quite prestigious to display these items in the store.

70

71

68 The three golden balls used by the Medici family during the Renaissance are still in use today.

69 The barber pole is a symbol that has survived hundreds of years.

70 The Cigar Store Indian, the most typical point-of-purchase display of the nineteenth century, is a collector's item today.

71 In the 1800s merchants created their own point-of-purchase sales aids.

Glass display windows—a major innovation—began to appear around 1840. The would-be consumer on the street could see and admire the new merchandise, cleverly and often profusely displayed in the window.

Point-of-purchase display has come a long way from the days of the pioneer merchants. But progress was very slow. For example, three-dimensional window displays did not appear until about 1900. Most store windows had the overstuffed look; merchants displayed in the windows large quantities of merchandise, to show variety. In the countryside, metal or painted signs on barns and blacksmith shops advertised products and services. Around 1910, die-cut, three-dimensional displays began to appear; by 1922 the neon sign became widely available. Today's point-of-

72

73

74

purchase display is one of the most important selling aids in retailing, and thus is crucial to the package designer's efforts. As merchandising, retailing, and selling techniques have become more and more sophisticated, so have the techniques of point-of-purchase display and packaging.

designing a POP display

To create an effective POP display, the designer must be knowledgeable in marketing and advertising procedures as well as in the physical layout of the retail establishment. As consumers, we are familiar with the layout of the supermarket, the drugstore, or the package (liquor) store, but AS DESIGNERS we must also understand retailing techniques for attracting the consumer and selling the product. It seems a simple

matter to design a display on the drawing board, but it is another matter to build it and to produce it within the required budget. This is no small undertaking.

Obviously, the POP designer is not just an artist who can use color, art, and typography. He or she must also have a total understanding of retailing problems in various types of retail establishments.

The first step in learning about retailing is a visit to a retail store. Four points in particular should be studied: the physical layout of the store; the flow of consumers; the placement strategy of various types of displays; and demographic and ethnic preferences in merchandise selection. Valuable information can be gathered directly from the store manager, who is usually a friendly source of information

72 Brand names started to appear on tins in the nineteenth century, as well as on barrels, crates, and glass.

73 Glass display windows, first appearing around 1840, were a major innovation for attracting consumers at the point of purchase.

74 Retail stores are valuable sources of information about packaging, displays, and consumer behavior. Photo courtesy of The Great Atlantic & Pacific Tea Company, Inc.

75 A turn-of-the-century premium, 1907.

74

75

and, as a person of responsible position, usually willing to hear as well as to suggest improvements and changes. A field trip is an essential step in the solution of POP problems. Companies as well as design schools should insist on regular tours to various retail outlets to learn about merchandising techniques of particular trades.

It would be most interesting and beneficial for the designer to study the manufacturing and packaging systems of products intended for use in the POP display. This study would give the right answer to such questions as: How should the product be packed and shipped? What are the problems of getting the product into the display? How should the display be shipped to the retailer? How should the assembly instruction sheet be prepared? These are all design-related problems that must be solved in order to create and design an effective POP display. Advertising, marketing, engineering, and production groups must be consulted in every phase of the design process. The design of the POP display, like any other three-dimensional graphic design, is a group effort. The designer must learn to function within the group.

The concept of the POP display is presented first in sketch form,

much like a package. Once the display concept is decided upon, an actual-size model is made. Since most cardboard displays are shipped unassembled, it is advisable to work out the precise assembly system for the display before building the final model.

The most important consideration affecting design is the budget. Many types of displays are possible, ranging from inexpensive, small shipper merchandisers made of E-flute corrugated to costly, permanent displays made of plastics, wood, and metal. With the new, improved flexographic printing on corrugated board, attractive, full-color displays can be produced at reasonable prices. Other types of corrugated displays— merchandisers, counter displays, floor stands, dump bins—can be designed with very effective graphic applications. The cost may range widely, depending upon the type of display, which can vary from simple flexographic printing to complex photographic effects where a full color pre-lithographed sheet is laminated to the board. The most popular and effective display is the merchandiser or promotion display. This versatile POP display can sell books, toys, cosmetics, housewares, practically all kinds of merchandise. The basic fact of

business is this. It takes money, time, and technical know-how to become a successful designer.

POP display categories

There are several types of POP displays, each serving a specific merchandising function. The major categories are display merchandisers, permanent displays, window displays, posters, signs, and vehicles. Each category has its own variations, depending on its purpose, location, and merchandising function.

The contemporary retail establishment is a busy place. A substantial percentage of selling is SELF-SERVICE, a system of selling whereby the consumers pick and choose the merchandise themselves, rather than having it brought to

them by salespersons. "Impulse buying" is directly attributed to displays called MERCHANDISERS. These are strategically placed in the store, with some small ones usually near the check-out counter. The merchandiser is sometimes called a PROMOTIONAL DISPLAY because it is designed for use only for the duration of a particular promotion. Structurally, the display merchandiser is built for easy

assembly. A well-known variation of the promotional display is the COUNTER DISPLAY, often called a shipper. This is generally a small display most popular with stores selling cosmetics, pharmaceuticals, novelties, and other small items. An interesting variation of the counter display is the GRAVITY-FED DISPLAY, which dispenses batteries, films, and similar objects. THE FLOOR STAND is a large structural display, used

77 Merchandisers designed for the duration of a particular promotion. Photo courtesy of Helena Rubinstein, Inc.

78 Merchandise can be shipped in its own display. Photo courtesy of Helena Rubinstein, Inc.

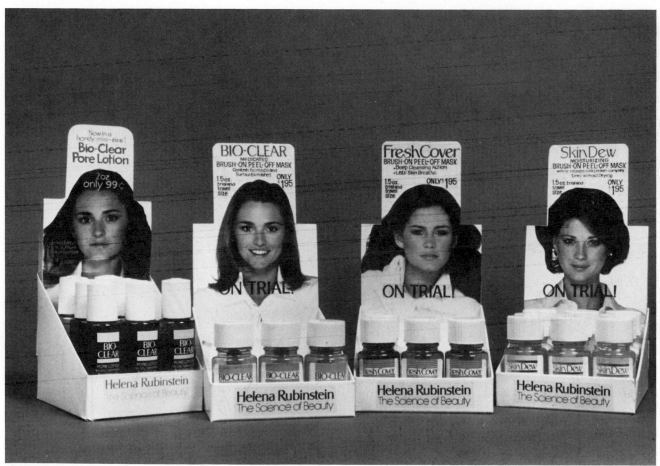

78

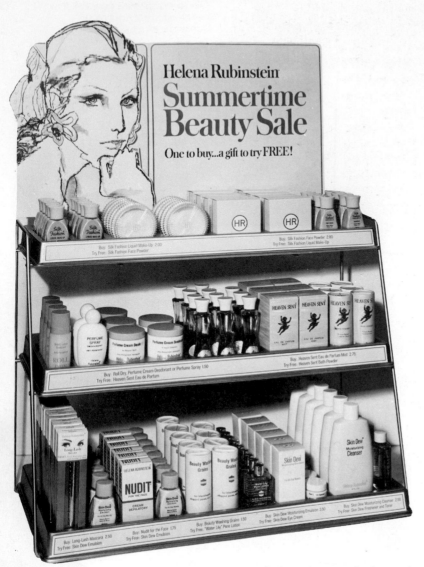

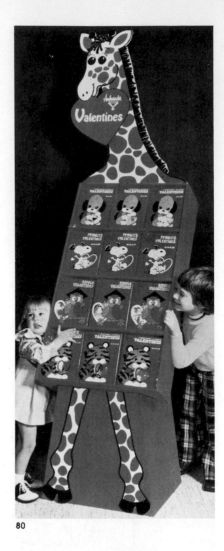

mostly in supermarkets and liquor stores, often animated with a small, battery-driven motor. THE DUMP BIN is a mass display unit into which merchandise can be "dumped" directly from the shipping carton to encourage consumers to "help themselves." This type of display is mostly used in supermarkets for sale-priced items. There are a number of other variations of displays, both in-store and outdoors (as at gas stations): mobiles, flags, banners, spinners, devices that move with or without electric power. The "send away" PREMIUM DISPLAY is another incentive POP device to stimulate sales. There are a number of other graphic devices like shelf-extenders, danglers, and signs to catch the attention of the consumer.

permanent displays represent an expensive category of elaborate

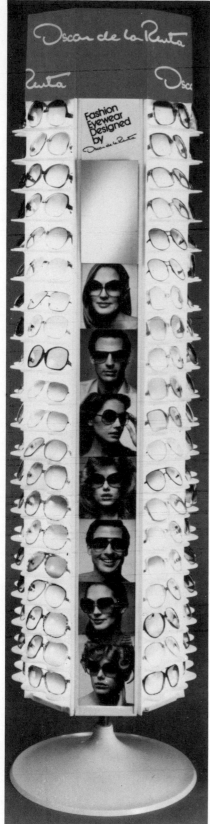

81

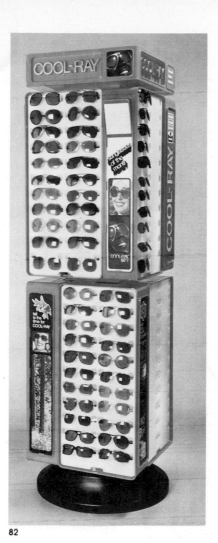

82

79 Counter displays can contain assorted products from the same manufacturer.

80 Floor stands are large, structural displays, used mostly in supermarkets and liquor stores. Photo courtesy of Hallmark Cards, Inc.

81 Counter displays and floor stands can be designed to revolve for maximum space utilization and attractiveness. Photo courtesy of Cool-Ray/Personal Care Brands—Division of Warner-Lambert Company.

82 Floor stands can be stacked to catch attention and for ease of movement, utility of space, and variation of design. Photograph courtesy of Cool-Ray/Personal Care Brands—Division of Warner-Lambert Company.

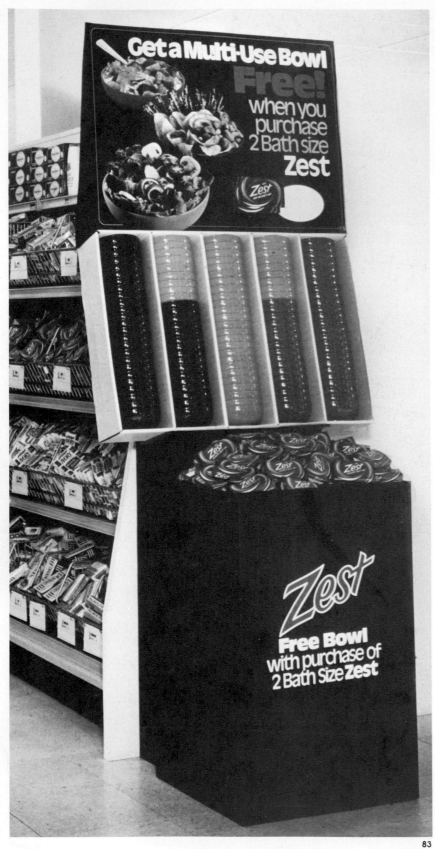

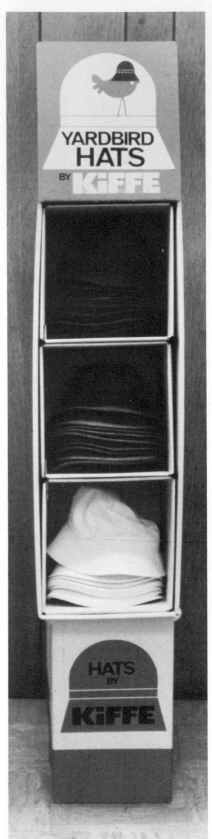

82

83   84

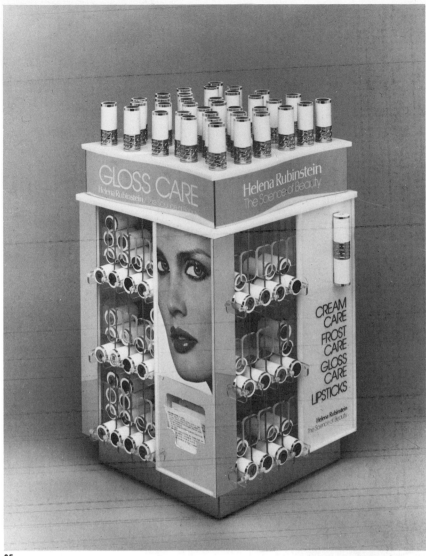

83 Dump displays can be designed to hold merchandise "dumped" directly from shipping cartons. Photo courtesy of Procter & Gamble.

84 Dump displays can also accommodate merchandise in neat stacks. Photo courtesy of Ideal Toy Corporation.

85 A permanent, gravity-fed display with sample testers above items for sale. Photo courtesy of Helena Rubinstein, Inc.

86 Another type of permanent display. Photo courtesy of Helena Rubinstein, Inc.

85

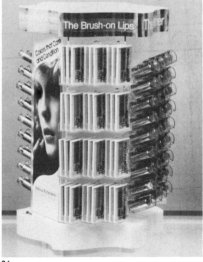

86

presenting

ENJOLI

The 8-Hour Perfume
for the 24-Hour Woman

ENJOLI

TESTER

compliments of Charles of the Ritz

88

displays, closely associated with the cosmetic and stationery trade. These are made of permanent materials: wood, plastic, or wire. Their function ranges from selling, to demonstrating (sometimes with the help of a specially designed computer), to helping the consumer to sample the merchandise.

window displays are probably the most valuable displays, for they can reach out to the consumer right in the street. A window display specially created by the store's

display department is NOT technically a POP display, although store-created window displays often include POP. The POP display has many variations, from a single simple but effective poster to the most intricate displays seen in the windows of travel agencies, camera and film stores, and liquor stores.

the poster itself is one of the oldest forms of promotional graphics. Some of the most prominent artists

87 Example of a permanent, contemporary POP display for a major soft drink. Reproduced with permission of The Coca-Cola Company.

88 Enjoli tester—a "give-away" counter display. Photo courtesy of Charles of the Ritz Group, Ltd.

89

90

91

89 A window display that makes use of its boundaries. Photo courtesy of Helena Rubinstein, Inc.

90 This poster, now a collector's item, was used to advertise an early twentieth-century movie.

91 One of the many posters created by Toulouse-Lautrec.

92 Signage from 1976 Olympic Games, Montreal. Design Hunter Straker Templeton, Ltd. Courtesy of Bicultural Information Committee of Public Works, Canada.

of this century, from Toulouse-Lautrec to Picasso, were involved directly in the creation of posters for theaters, ballets, art galleries, and political propaganda. The art of poster design is a highly creative form of expression and can be a powerful political weapon as well as a work of art.

signage is a point-of-purchase graphic which identifies companies and products and which may carry an important message. A typical example of this type of POP is found in fast food outlets and gas stations, where a TOTAL ENVIRONMENT may be created with signs. Automobile dealerships and appliance makers' showrooms, international fairs, and even the Olympic Games are other excellent examples of the use of effective signage as a POP.

Major soft drink, beer, and camera manufacturers distribute wall plaques, clocks, and indoor and outdoor signs and neons for restaurants, bars, and retail stores to advertise their products. This is a subtle way of incorporating POP as an advertising device, as well as providing a service medium for the consumer.

The **commercial vehicle** is another great selling device. The circus wagon, the simple ice cream cart, and the truck bearing a company's name or logo are moving billboards of POP advertising. We live in a mobile society. There are

92

93

mobile amusement rides, mobile libraries, mobile medical treatment and diagnostic vehicles, and, in our larger cities, even mobile theaters.

suggested reading

Offenhartz, Harvey, POINT OF PURCHASE DESIGN. New York: Van Nostrand Reinhold, 1968.

94

95

project

First, make a rough sketch for a POP display for a small counter display or dump bin of your choice of merchandise. (Suggestions: paperback books, toys, candies, cosmetics.) Work out your structural problems in sketch form (see illustrations in chapter). Now construct the display from white corrugated board or cardboard. For the dump bin, build a scale model (the average dump bin is about six to seven feet tall). Use proper lettering (copy) and illustration. The display must also serve as a shipper for the merchandise. Unassembled, it should make a rectangular shipping carton.

CHAPTER FOUR

plastics primer

What you should know about plastics. An introduction to the world of plastics. Complete, up-to-date information on different kinds of plastics. Where plastics come from. What plastics are used for. Everything and anything a designer should know about plastic technology. Methods of molding, shaping, and fabricating. Various packaging mediums: the thermoformed package, blister pack, skin pack, rigid and semirigid containers. An alphabetical plastics glossary for easy reference.

We are living in the age of plastics. Look around you. The chair you are sitting on probably has a plastic fabric covering, maybe even plastic parts. You may have a plastic-coated desk. Your car has a plastic dashboard, plastic upholstery, and the engine has many plastic parts. In the office, you drink your morning coffee from a disposable plastic cup; you may have bought your dinner from the supermarket in an airtight, heatsealed plastic bag or tray. Even your frying pan has a non-stick plastic coating. The chances are that you are wearing some articles of clothing made of plastics, probably a "wash and wear" shirt or dress; your shoes contain "man-made materials." If you had some recent dental work done, you may have plastic in or on your teeth. If you wear glasses or contact lenses, these could be plastic too. Plastics are all around us.

One thing quickly becomes evident—there are a great many kinds of plastics. In your home there are at least thirty different kinds of plastics. Some are hard, some are soft, some are flexible, some are transparent. Each plastic has its own characteristics that make it suitable for its particular use or function.

When we think of "plastics," we usually think of man-made materials or substances, but the fact is that a great variety of plastic substances are to be found in nature. Beautiful AMBER comes from trees, as does ROSIN. LAC is a substance made by an insect, which we make into a commercial product called SHELLAC. RUBBER is another kind of natural plastic that comes from the rubber tree. CASEIN comes from milk. Plaster of Paris is another byproduct of nature, as is cement.

The ancient Egyptians and Romans used resin and bitumen for sealing documents. This tar-like substance was found all over the world. The famous Pitch Lake in Trinidad was discovered by the great English explorer Sir Walter Raleigh in 1595. The sticky substance was used for centuries to waterproof ships, boats, and water tanks. When Columbus visited the Americas, he saw Indians playing with bouncing balls made from the gum of a tree. In 1924, the Scottish chemist Charles Macintosh began manufacturing waterproof raincoats made from the same rubber. His "macintoshes" became world famous. The great discovery that transformed the rubber industry was the vulcanization process, invented by the American Charles Goodyear, which strengthens the rubber and makes it heat- and cold-resistant.

chemistry for the designer

More than a hundred years ago, chemists began searching for substitutes for ivory, a scarce and expensive material. An English chemist, Alexander Parkes, found that if cotton was treated with acid and camphor a tough, bone-like substance was formed. He called this substance PARKESINE. It was exhibited at the Great Exhibition of 1862, but it was a total failure. In 1829, two American brothers, John and Isaiah Hyatt, were more successful with a new product called CELLULOID, the first man-made plastic.

The science of man-made plastics dates back to the nineteenth century. During the mid 1800s, extensive studies were made in chemistry, particularly on solids which formed mysteriously from cellulose, rubber, silk, and some natural substances.

Scientists knew that every substance is made up of tiny units called MOLECULES. Each molecule is itself made up of particles, or ATOMS, of one or more chemical elements. Most substances like water have short molecules, containing only a few atoms. Plastics have molecules that may contain millions of atoms linked together so that they form a chain.

The simplest and best known plastic is polyethylene. This plastic is made up of atoms of carbon (C) and hydrogen (H) which are linked together like this:

$$\begin{array}{ccccccccc} H & H & H & H & H & H & H & H \\ | & | & | & | & | & | & | & | \\ \cdot C \cdot & \cdot C \cdot & \cdot C \cdot & \cdot C \cdot & \cdot C \cdot & \cdot C \cdot & \cdot C \cdot & \cdot C \cdot \\ | & | & | & | & | & | & | & | \\ H & H & H & H & H & H & H & H \end{array}$$

Notice that the molecule is made of many repeated parts or units, consisting of two hydrogen atoms to every one of carbon (CH_2). A substance with a long chain of molecules is called a POLYMER, a Greek word meaning "many parts." (One MER is called a MONOMER.) The process of joining these molecules together is called POLYMERIZING. When monomers of different molecules are chemically joined together (polymerized), the resulting compound is a COPOLYMER. If they are mixed together in polymer form, the result is a BLEND. The names of plastics often begin with POLY to describe the structure. The last part of the word indicates what the plastic is derived from (for example, ethylene). Thus we can tell the origins of plastics like polystyrene, polyester, polyvinyl-chloride, and so forth. Sometimes plastics are better known by their trade names. For example, polyamides are the familiar Nylon (by DuPont); polyesters are labeled Dacron or Fortrel; acrylics are marketed as Acrilan, Dynel, or Orlon. Names of plastics are often abbreviated, like PVC, which is polyvinyl-chloride.

Plastics are chemically synthesized from crude oil, coal, natural gas, salt, air, water, and certain agricultural products such as cotton, soybeans, and trees. They are divided into two groups: the THERMOPLASTICS (polyethylene, polystyrene), which can be softened or recycled by heat; and the THERMOSETS (phenolic, urea, melamine) which, once molded, cannot be softened for further processing or recycling.

Chemical engineers can create and build many desirable features and combinations into plastic polymers and copolymers. Plastic RESINS, the raw materials for plastic moldings, are usually produced by chemical companies such as Dow, Monsanto, DuPont, American Cyanamid, and others. Since crude oil is the natural source of most plastics, the major oil companies have also become involved in the manufacture of plastic resins.

Plastics can be rigid, flexible, transparent, opaque. They can be made in any specified color, including fluorescents. Plastic can be made into sheets, rods, tubes, fabrics, films, and foams. Plastics can be MOLDED, EXTRUDED, SHAPED, and FABRICATED.

molding processes

Molding is one of the simple ways of shaping plastics into objects. Molds for plastic molding are made of metal by expert tool- and die-makers, directly from a three-dimensional model made by the model-maker and based on the designer's rendering. There are several methods of molding, depending on the item to be made and the materials used.

injection molding is used only for THERMOPLASTICS. The plastic resin (granules or powders) is first melted in a heating chamber and then forced by a plunger into an injection-molding machine. When the plastic has set in the mold, the mold opens and the product is ejected. The mold closes again and the process is repeated. A wide variety of objects can be made quickly and inexpensively in large quantities by this method. Housewares, toys, boxes, and components are typical injection-molded products.

96 Nineteenth-century chocolate molds.

97 These toys were made with an injection-molding machine.

96

97

blow molding is similar to glass blowing. Air is blown into a blob of molten plastic inside a mold. The plastic expands until it reaches the walls of the mold. It takes the shape of the mold and sets. The mold then opens and the object is ejected or removed. Blow molding is used for making plastic bottles and large, hollow objects.

rotation molding (sometimes called slush molding) is another way of making hollow objects from thermoplastics. Plastic resin is placed in a hollow mold, which is then rotated. The inside becomes covered with the plastic coating, which hardens to form the object. Dolls' heads and highly detailed objects are made with this method.

compression molding is used for THERMOSETTING plastics only. Thermosetting plastics cannot be molded by other methods, because they melt and set at the same time. In compression molding, the thermosetting powder is placed in the bottom half of a HEATED mold. As the powder melts and begins to set,

100

the upper half of the mold descends and forces it into shape.

extrusion is the process by which rods, tubes, pipes, and film are made. The method is simple. Molten plastic or plastic granules are fed into the heated extrudes. A screw device forces the plastic through a shaped hole (a die) and the product takes the shape of the hole. To make rods, the die is a simple hole. To make tubing, the plastic is pressed between the die and a rod inside to create a hollow tube. Variations of this

method are used for coating wires, paper, and other sheet materials.

welding is a method that involves heating the edges of two pieces of plastic and then joining them together. The method is similar to metal welding. Another recent method is HIGH-FREQUENCY or SPIN WELDING, whereby one piece of plastic is spun very rapidly against the other. This generates heat at the joint and fuses the two parts together. Tubular containers are usually welded together.

98 These hollow containers were made by blow molding.

99 Welded plastic bottle. Courtesy of Chester Plastics Ltd.

100 These products were thermoformed. Courtesy of Creative Packaging Inc., Packaging Systems Corporation, and The Great Atlantic and Pacific Tea Company, Inc.

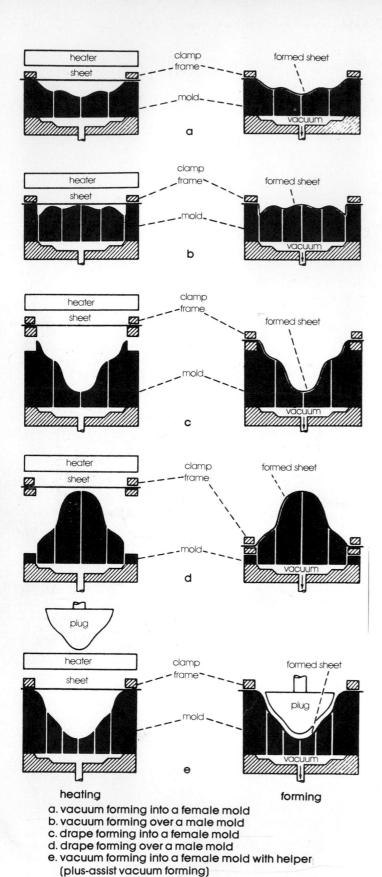

heating | forming

a. vacuum forming into a female mold
b. vacuum forming over a male mold
c. drape forming into a female mold
d. drape forming over a male mold
e. vacuum forming into a female mold with helper
   (plus-assist vacuum forming)

thermoforming means heat shaping, that is, the shaping of the heated thermoplastic sheet or film through forced contact with a mold. Because a vacuum is often used in forming the sheet part, the process is also called VACUUM FORMING.

Like most molding or shaping processes, thermoforming is a simple process. There are several variations of thermoforming systems available depending on need and the type of package and plastic used. Some of the variations are shown to the left.

what you see is what you get

In addition to the standard thermoformed packages, there is a wide assortment of cups, trays, lids, and special shapes made by thermoforming. Thermoformed packages, where the product is sealed between a transparent bubble or sheet and a printed card, have become a major packaging form, because they have proved successful in selling more goods faster in self-service retailing. In addition,

thermoformed pharmaceutical, medical, and surgical packages and contents can be sterilized by gas, gamma radiation, or Cobalt 60 as part of the packaging process. This process is an important milestone in medicine made possible by packaging technology.

Since thermoformed packages represent a large group of

101 Various techniques of thermoforming.

102 Basic thermoformed card pack constructions.

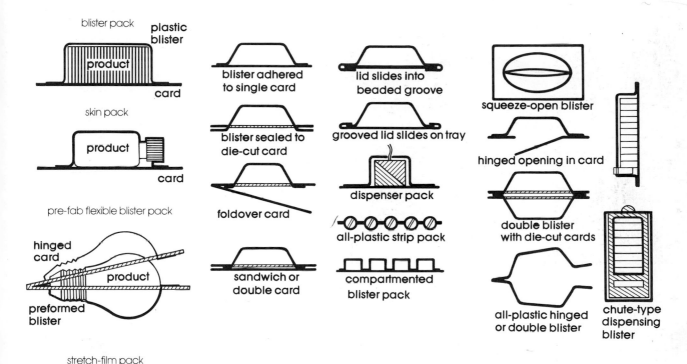

102

packaging applications, a thorough knowledge and understanding of the system is absolutely necessary for the beginning designer. The following definitions should help to clarify the terminology that has specific meaning in thermoforming.

blister pack. Primarily a visibility pack, this is the most widely used of all card packs. It is used for all types of products that can be affixed to a card. Usually acetate, polystyrene, vinyl, or similar types of sheet plastics are used for the blister.

prefabricated blister and card. A preformed blister is attached to a die-cut, hinged window card to fit the item inside the bubble. The item is visible on both sides of the card. This is used for bulky objects, hardware, some cosmetics, and notions. Economical films such as polyethylene and vinyl are used.

skin pack. The product is used as the mold or die. It is laminated (sealed) to the card with a thin but tough film, mostly acetate, polyethylene, butyrate, or vinyl.

stretch film pack. A vinyl film is bonded to a die-cut double window card to encase the product in a see-through visual display card that combines the features of blisters and skin packs. It is used for cosmetics, bottles, tubes, hardware, and electrical components.

Thermoforming offers many options for the designer. But as can be seen from the diagrams, many factors must be taken into account when planning and developing a thermoformed package. In the hands of the creative and informed designer, this is one of the most effective packaging formats.

In certain areas, thermoforming almost replaced the MOLDED PULP packaging materials. In the early 1900s, Martin Keyes invented the first machine to mold paper plates from wood pulp. By the 1930s, industrial packaging included a great variety of shapes for packing eggs, glass, and other fragile or perishable goods. Molded pulp packaging is inexpensive because of the low material cost. However, a large quantity of a particular product must be run to take care of the initial die costs.

projects

The first plastic package you will be asked to design will probably be a blister package or a skin pack. After reviewing the detailed descriptions and diagrams in this chapter, you should be able to come up with a winner. For a blister package, use the existing blister from a product card. Redesign the entire card—new shape, new graphics, and a brand new presentation are required.

Create a brand new STRETCH FILM PACK. Use any of your

favorite products, but devise your own card and graphics. Use a Saran type of film. You can stretch the film with the heat of a hairdryer. Once the card is pasted together with the product, you can start "shrinking" the film with the dryer. This method is most suitable for laminating objects to cards and for covering photos, artwork, and three-dimensional comps.

You can make boxes, cabinets, models, and even gifts with this plastic. For tools, all you need is a mat knife, some fine sandpaper, and a steel ruler. You also need polystyrene sheets and Rez-N-Bond, available at plastic sheet supply houses or occasionally at art supply stores. Polystyrene usually comes in large 40" × 72" sheets and is surprisingly inexpensive. Thickness varies from .020-.250 points (I prefer to work with the .050-.060 thickness). Polystyrene comes in opaque white and can be painted or sprayed with acrylic colors. The great advantage of this material is that it can be adhered with the Rez-N-Bond within seconds. The bond is non-toxic and non-flammable, and is also suitable for adhering Lucite or Plexiglas. Polystyrene can be scored with a mat knife, and will snap away easily. Virtually any type of box or container can be made with this material. Your first project should be a simple box with a separate lid. Once you are experienced working with this new material, there is no limit to what you can build or construct with it. Polystyrene sheets and the recommended bond are used by the most professional model makers.

suggested reading

Griffin, Roger C., Jr. and Stanley Sacharow, PRINCIPLES OF PACKAGE DEVELOPMENT. Westport, Ct.: The Avi Publishing Co, 1972.

MODERN PACKAGING ENCYCLOPEDIA.

PACKAGING DIGEST, 1976.

PACKAGE ENGINEERING, 1978, 1979.

Packaging Institute USA, 1979.

Packaging Systems Corporation: THERMOFORMING. Orangeburg, N.Y.: 1978.

Sacharow, Stanley, and Roger C. Griffin, Jr., BASIC GUIDE TO PLASTICS IN PACKAGING. Boston: CBI Publishing Co., 1973.

rigid and semirigid plastic

Since the early 1950s, rigid and semirigid plastic containers have made major inroads into virtually every product category. Plastics are preferred materials for packaging. Versatility, resistance to breakage, and light weight are the chief characteristics of the plastic container as a packaging format.

There are several basic packaging forms in the rigid and semirigid category of packages. These are: trays and platforms; fabricated, molded, or thermoformed boxes; plastic bottles, jars, vials, pails, and carboys; and plastic cans, crates, barrels, and drums.

trays and platforms

A tray is a lidless container for carrying heavy objects. A tray can be a simple divided platform like those in candy boxes, or it can be designed for use in various shapes as meat or produce trays that are sealed in plastic film. Plastic trays and platforms can be thermoformed or injection molded from a variety of plastics including polyethylene, polypropylene, polystyrene, PVC, acetate, or expanded polystyrene. Since almost all trays are disposable, they are made from the least expensive plastics, with the least expensive method, usually thermoforming.

103

104

boxes

The early plastic box was a fabricated container, usually of acetate or Plexiglas, joined with cements or solvents. Today, most plastic boxes are injection molded for greater versatility in design, and they are often combined with thermoformed platforms. These boxes are mostly used for cosmetics and other luxury items. Injection molded boxes come in stock sizes in a great variety of styles. Like all plastics, these can be decorated by printing, hot stamping, screening, or labeling. One of the commonly used plastics is polystyrene, a low-cost, clear or opaque plastic. Occasionally polypropylene is used when better hinges are required.

103 Clear plastic tray for produce. Photo courtesy of The Great Atlantic and Pacific Tea Company, Inc.

104 Plastic tray sealed in clear plastic film. Photo courtesy of The Great Atlantic and Pacific Tea Company, Inc.

105 An assortment of injection-molded boxes. Photo courtesy of Bradley Industries.

105

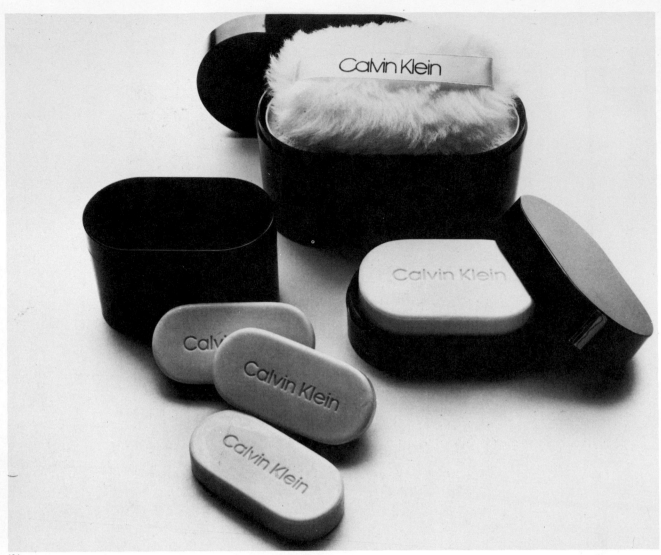

containers

Plastic has certain definite advantages over glass. The fragility of glass containers has been a strong selling point for plastic containers. Now that most product compatibility problems in the detergent and personal product markets have been solved by using compatible plastic materials for the containers, most liquid detergents are now sold in plastic containers, and more and more toiletries are being packaged in plastic bottles. Plastic bottles are slowly infiltrating the $90 billion food and beverage industry. New resins suitable for bottle molding appear constantly. Sohio's Barex, Monsanto's Lopac, American Cyanamid's XT Polymer, and Richardson's NAS and SAN polymers are some of the contenders to capture this lucrative but highly competitive market. The traditional plastics for bottles are PVC and polyethylene (the first "squeeze" bottles of the 1950s). Polypropylene, polyester, and polystyrene (from the late 1950s and 1960s) are still being improved to capture new markets.

The plastic industry offers the designer new possibilities to create unusual yet functional bottles and closures for the ever

increasing demands of the marketers. Beginning designers should study the new plastic resins, their unique function and product compatibilities. These can open up new areas in which to originate concepts in bottle and container design. The versatility and almost unlimited design possibilities of the plastics is probably the most challenging aspect of being a designer today.

The procedure of designing plastic bottles and jars is similar to that of glass bottles and jars. However, the designer working with plastics may have a greater variety of materials and decoration processes. Keep in touch with your supplier about recent developments in new materials and processes. The development of blow-molding equipment made the plastic bottle a practical, inexpensive package. As with glass, a variety

106 Red lacquered oval boxes. Photo courtesy of Calvin Klein Cosmetics, Inc.

107 This plastic bottle is packaged in a folding carton, a typical cosmetic industry format. Photo courtesy of Jean Naté/Charles of the Ritz Group Ltd.

of stock bottles and closures is available. They present interesting redesign problems for the designer.

plastic vials are small bottles usually injection molded or fabricated from tubing. Vials are used to package small items, pills, capsules, or powders.

pails are large plastic containers (garbage cans, trash bins) usually molded from high-density polyethylene.

carboys are large three- to thirteen-gallon plastic bottles used for commercial liquid products and usually encased in rigid outer containers. Large half-gallon and gallon bottles for milk and alcoholic beverages are making strong headway.

plastic cans are slowly infiltrating the market in the United States. An interesting development in the frozen beverage field is the "total" plastic can. Recently in Europe, flexible or rigid plastic containers for petroleum products as well as other consumer goods have been introduced. These plastics are relatively inexpensive compared with the tinplate cans still used in this country.

uses in shipping

Crates and shipping containers used to be made of wood. Barrels and drums were made from bent staves held together by hoops. The disadvantages of conventional materials are their weight and their tendencies toward corrosion and other forms of deterioration. Since most of these packages are costly, they are generally made for expensive products.

Plastic shipping containers have made inroads into the traditional shipping crate market. Plastic crates are now used for carrying milk, soft drink, and beer bottles. Plastic drums offer greater versatility and lower cost than the traditional wood or fiber barrel. Pallets are rigid platforms, usually made of lumber, for stacking cartons, bags, or sacks. A variety of plastic pallets are now being marketed; some are injection molded, while others are made of lightweight, expanded polystyrene foam.

108

a plastics glossary

An infinite variety of combinations can be found within each family of plastics. It would be impossible to cover all available plastics in this chapter, but we have covered some basics. Research by plastic and chemical manufacturers is aimed at finding combinations suitable for specific purposes. Since plastics are man-made, literally thousands of molecular variations are possible. New arrivals add to the designer's repertoire. Some plastics become favorite commercial items; others disappear rapidly when it is found that another combination does a better job at lower cost.

While suitable plastics for the designer are usually specified by engineering, the designer must have an understanding and working knowledge of the various plastics and their characteristics as they are used for products and packaging.

abs resins Rigid, strong plastics. Most suitable for injection-molded boxes, containers, trays, and closures. Excellent for many foods and drugs. Very low cost.

acrylics A large family of acrylic plastics: the rigid plastic sheet Plexiglas, the extruded Lucite, and the famous acrylic fibers appearing under trade names such as Orlon, Acrilan, and Dynel® Acrylic paints are very versatile.

acrylonitrile polymer A recent arrival of the acrylic family. Used for soft-drink bottles. Transparent, lightweight, most suitable for blow molding. Comes under the commercial names of Barex and Lopac.

bakelite The first plastic made entirely from chemicals was named after Dr. Leo Baekeland, a Belgian chemist who first made it in 1909. Bakelite is a thermosetting formaldehyde plastic. It is used for heatproof handles on pans, ashtrays, and electric iron components, and has many other uses where heat resistance is required.

cellophane First made from cellulose by Jacques Brandenburger in Switzerland in 1908. Production of cellophane in the United States began in 1924. About 350 million pounds of cellophane are used annually by the food industry.

celluloid The first man-made plastic. At one time it was widely used for men's shirt collars. In World War I it was used to coat the fabric on airplane wings. The first movie film was made of celluloid. Its great advantage is that it is flammable. Other plastics are now replacing it.

cellulose acetate A plastic everyone knows as a fabric called acetate RAYON. As a crystal clear film, it is used to make photographic film and "windows" for display cartons. In solid form, cellulose acetate is tough and has a high melting point. It can be blow molded or injection molded.

epoxy resins Thermosetting plastics that make powerful adhesives. They come in two tubes, one containing the resin and the other a chemical catalyst to harden the resin. Epoxy resins are used for coating and can be cast into shape by low-pressure molding.

foams Flexible plastic foams are made of a plastic called POLYURETHANE. They are made by mixing the plastic with a foaming agent to produce bubbles of gas, which become trapped to form a foam. Polyurethane is used for lining, carpets, mattresses, pillows, sponges, and many other items. Some rigid foams are also manufactured. The best known is made from POLYSTYRENE and is called "expanded" polystyrene. It is used as a packaging material for appliances, cameras, typewriters, and other rigid but delicate products. It also makes an excellent insulating material.

melamine Another thermoset plastic. A well-known material for plastic dishes.

phenol formaldehyde A thermoset plastic. Excellent resistance to alcohols, superior heat resistance. For closures, tubes, jars, and bottles.

polyamide Better known as Nylon. Introduced by DuPont in 1938 as a fiber for woven or knitted fabrics. Nylon became a commercial packaging film in the late 1950s. Nylon film is used for cook-in pouches and to protect grease- and oil-bearing foods and pharmaceuticals.

polyester The polyester family is best known in the form of synthetic fibers, sometimes marketed under the trade names of Dacron and Fortrel. Polyester fabrics keep their shape, because the fibers do not stretch.

polyethylene The most familiar of all our plastics was discovered in England shortly before World War II. Practically everything we buy these days is wrapped in a "poly" bag. Polyethylene is widely used as a film for packaging, because it is inexpensive, transparent, and tough. Jars, bottles, tubes, and drums are molded from polyethylene. It is available in low, medium, and high density, depending on need. About 1.5 billion pounds are sold annually in the United States.

polypropylene Discovered in 1954 by Professor Guilio Nata in Milan, Italy. It can be molded and is also available in film form. Excellent for bottles, jars, closures, thermoformed packages, and components.

polystyrene This plastic has an early history. It was first synthesized in 1866. During World War II the need for synthetic rubber led to the development of styrene rubber (Buna S). Polystyrene is the most versatile plastic. It is widely used for packaging in film as well as in solid form. Especially known for injection-molded clear boxes, cups, closures, and trays.

polytetrafluoroethylene Commonly known as Teflon. A member of the fluorocarbon family. Teflon is extremely resistant to heat and corrosive solvents. Can be molded and produced in film or fiber form.

polyvinyl chloride (PVC) and polyvinylidene chloride (PVDC) Developed during World War II. First introduced as a commercial packaging material in 1946 by the Dow Chemical Company. The popular Saran type of packaging film and the colorful vinyl, a leather-like plastic, are PVC plastics. Excellent for blow-molded bottles, thermoformed cups and trays. It has a glass-like appearance and is most suitable for cosmetic and pharmaceutical bottles. In extruded form, PVC makes excellent piping, garden hoses, and electrical wire and cable coverings.

rubber hydrochloride Developed by Goodyear in 1934. The first non-cellulosic, transparent, thermoplastic film. Named Pliofilm by Goodyear. This film is used as a heat-sealing medium for flexible packaging, especially for meat and dairy products.

urea Another thermoset plastic. Has high rigidity and strong chemical resistance. Used for closures and boxes for luxury items.

water-soluble plastics Used in pouch packaging for bath products, shampoos, rinses, dyes, food mixes, and many other household and industrial products. This type of package eliminates product waste and gives accurate dosage or measure. The future of the water-soluble films are most challenging for the designer. They offer convenience and economy in packaging various mixes and ingredients. As an interesting project, develop some NEW applications for this exciting medium. Do some "in-store" research first.

CHAPTER FIVE

flexible

packaging

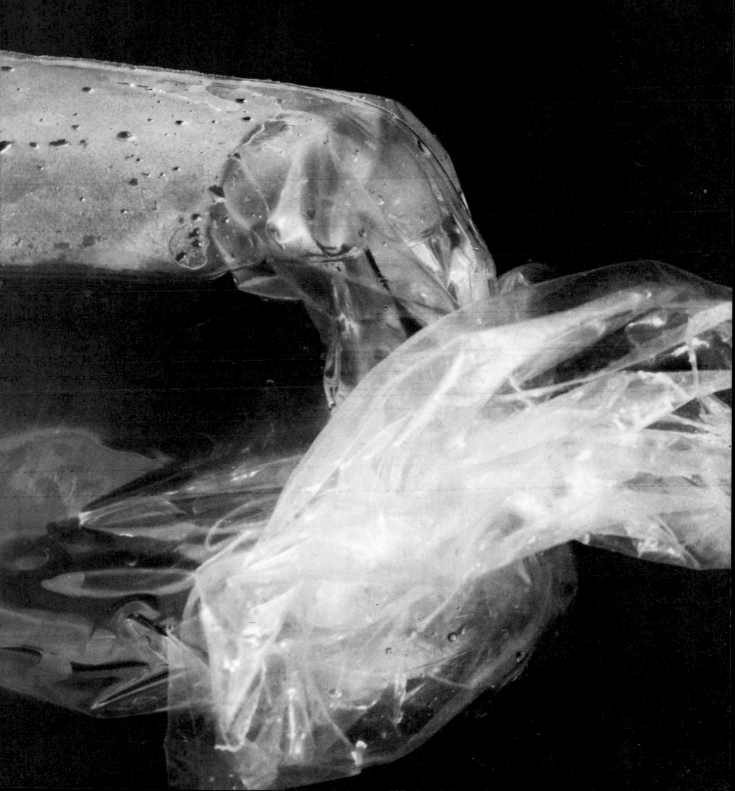

flexible packaging

Bags, wraps, and pouches. A detailed description of the flexible packaging of today and of the future. A great variety of bags and pouches, including space-age packaging. A special section, designing shopping bags, includes a design project.

Bags and wraps made from woven plants and animal skins were probably some of the oldest forms of flexible packaging used. Flexible paper packaging was one of the byproducts of our Civil War. Cotton bags had been previously used for the storage and transportation of food. After hostilities broke out, and Northern producers were cut off from the source of cotton, paper became the logical substitute. The mass-produced grocer's bag and paper sack were created by George West, a New York mill owner, whose Civil War paper sacks were the ancestors of our modern MULTIWALL bags.

The introduction of cellophane and the development of plastic films, foils, and specialty papers, as well as methods of coating and laminating, created new types of flexible packaging materials. The largest user of flexible materials is the food industry, which uses about 80 percent. Some of these flexibles designed to freeze, cook, mix, preserve, and dispense are having a dramatic impact on the food industry, especially in convenience foods.

There are three basic types of flexible packages: wraps and overwraps, preformed bags and envelopes, and form-fill-seal pouches.

109 The largest user of flexible packaging materials is the food industry. Photo courtesy of The Great Atlantic and Pacific Tea Company, Inc.

109

wraps and overwraps

110 Basic types of flexible packages. Photo courtesy of Procter & Gamble and The Great Atlantic and Pacific Tea Company, Inc.

111 Types and uses of bags and envelopes.

The WRAP is a sheet of flexible material, usually fed from a roll stock and formed around a product such as a candy bar or a loaf of bread. The OVERWRAP is formed around a basic package such as a carton. Variations of the wrap are bands and sleeves. Specialized wrapping machines have been developed with very high-speed performance and automation. Closures for wraps may be adhesives, heatseals, or peelable closures.

110

preformed bags and pouches

The preformed bag is basically a tubular construction, fabricated from paper, plastic, foil, fabric, or a combination of these materials. There are four standard styles of paper bags: flat, square, self-opening style (SOS), and satchel.

flat bag. Invented about 1850, this is the oldest form of paper bag. Made simply by folding paper into a tube, it is used for hosiery and textiles.

square bag. A bag with GUSSETS (pleats), designed to hold bulky materials. Originally made of paper, it is now available in high-density polyethylene for industrial products.

self-opening bag (SOS). Has side gussets and built-in flat bottom. The popular grocery and shopping bags belong to this group.

satchel-bottom bag. Like a flat bag, it has no gussets in the sides, but it has a flat base. It is used for bulky materials, mostly food, especially coffee and flour.

In addition to these styles, there are a series of multiwall bags, designed for heavy service (cement, plaster, fertilizers, chemicals). A variety of closures are available for bags: sealers, twist-ties, plastic clips, coffee tabs, and drawstrings.

paper styles

flat

Has lengthwise back seam, no gussets. Bottom is generally folded over and pasted or heat sealed. Simplest bag style, most economical in use of materials.
Typical closures: Heat seal, folded and pasted, tape, tie, clip, staple.
Some uses: Potato chips, snack foods, ice-cream bars, soft goods, powered foods, coffee, frozen foods, hardware, machine parts.
Some features: Rack display at point of sale; billboard label area; fewer folds lessen water-vapor permeability.

automatic

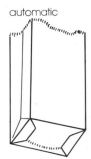

Self-opening style—SOS. Has side gussets, built-in flat bottom. It is rectangular in shape and self-standing.
Typical closures: Heat seal, folded and pasted, tape, tie, sewn.
Some uses: Coffee, cookies, dried milk, candies, insecticides, plant foods.
Some features: Has the best stacking ability, is easiest to handle. Stand-up display. Opens and fills easily.

square

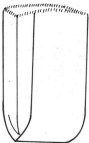

Has side gussets (also called tucks or bellows folds). Bottom turned up and pasted or heat sealed.
Typical closures: Heat seal, folded and pasted, tape, tie, clip, staple.
Some uses: Fresh produce, snack foods, soft goods, candies, cookies, plant foods, insecticides, cereals.
Some features: Cubic capacity is greater than in non-gusseted bags of similar width. Side folds and bottom may be printed for greater display value no matter how bag is stacked.
111

satchel

Has flat body, with no gussets. Bottom is formed in hexagonal shape, is flat and self-standing when filled.
Typical closures: Heat seal, folded and pasted, tape, tie, sewn, clip, staple.
Some uses: Sugar, flour, dried foods, cookies, baked goods, candles.
Some features: Combines economy of material with ease of handling and stacking. Stand-up display. Opens and fills easily.

film styles

bottom gusset

Has bottom gusset and side welds. Lip is optional, can be die cut with holes for wicketing. Made of thermoplastics, generally polyethylene.
Typical closures: Twist tie, plastic clip, tape, heat seal.
Some uses: Bread, paper goods, toys, multiple packages, household items, hardware.
Some features: Suitable for high-speed, automatic filling on newly developed bagging machines. Easy opening, reclosable, reusable. Offers visibility, strength, large-volume capacity.

flat-wicketed

Has side welds, no bottom seam or gusset. Made with or without lip. Lip can be die cut with holes for wicketing. Made generally of thermoplastics.
Typical closures: Twist tie, plastic clip, tape, heat seal.
Some uses: Produce, hosiery, candies, snack foods, hardware, toys, fresh meats.
Some features: Simple construction, visibility, easy opening, reclosable, strong, reusable.

specialty styles

contour

Two-compartment, foldover construction; suited to packaging multiple units, with advantage of enabling product inspection.

wallet

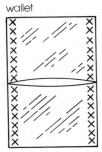

Fits rounded or irregular products; often made of shrink film or polyethylene. Used for phonograph records, poultry, fresh meats.

roll-fed

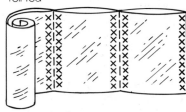

Thermoplastic film, often polyethylene, is perforated between bags for easy separation. Used in small-volume hand-filling operations.

bag closures

hood style

drawstring

coffeetab fold

plastic clip

plastic clip

twist tie

zip fastener

112

form-fill-seal pouches and variations

This was the first totally automated package form. It comes in roll form, and is filled and sealed on high-speed packaging equipment. With its endless variety of styles and geometric shapes, the pouch has been responsible for the development of many new products that could not have come into being were it not for this package form. Typical examples are the boil-in pouch for frozen foods, instant soups, coffee, and gravies; alcohol or fragrance dabs; instant shoe shines; shampoos; and packs for pharmaceuticals. Pouches for liquids are not new, of course; wineskins were used in biblical times.

A recent development in flexible packaging, originally developed in Europe, is the packaging of fresh milk in a tetrahedral (pyramid-shaped) pouch. Milk packaged in these containers has a shelf life of six to eight weeks. Wine, motor oil, and soft drinks are now being marketed in flexible packages.

Space-age packaging, originally designed for the Apollo flights, uses plastic laminated pouches. Sterilized, snap-open polyethylene pouches are used for various medical kits; surgical gloves, gowns, and other hospital disposables usually come in a variety of pouches which enable hospitals to cut time and labor.

112 Types and uses of bags and envelopes.

113 Flexible packaging of liquid products is a recent development.

113

114 A display of Apollo space food for three typical meals. Photo courtesy of National Aeronautics and Space Administration.

115 Under zero-gravity conditions, Velcro tapes and springs are used to hold this variety of packages in place. Photo courtesy of National Aeronautics and Space Administration.

114

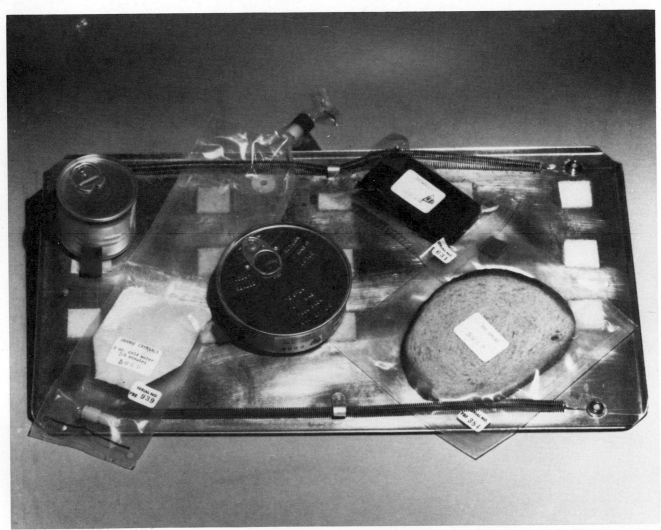

115

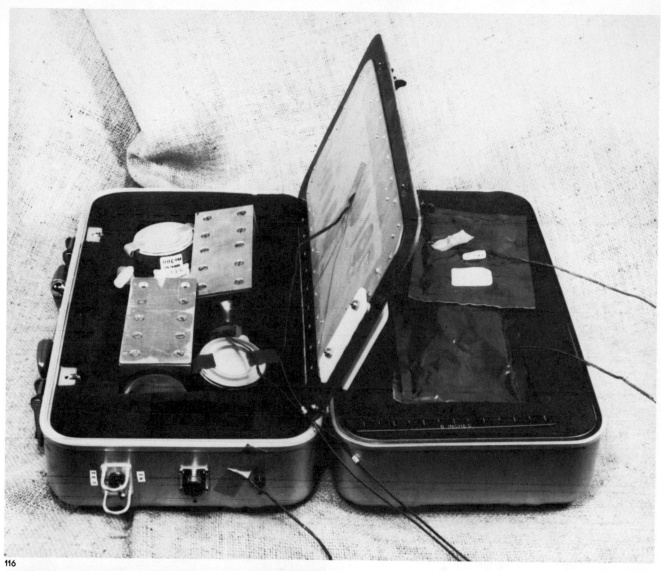

116

117

116 This carry-on.
plug-in equipment is used to
heat meals in space.
Photo courtesy of National
Aeronautics and Space
Administration.

117 A tested and proven package
for space food.
Photo courtesy of National
Aeronautics and Space
Administration.

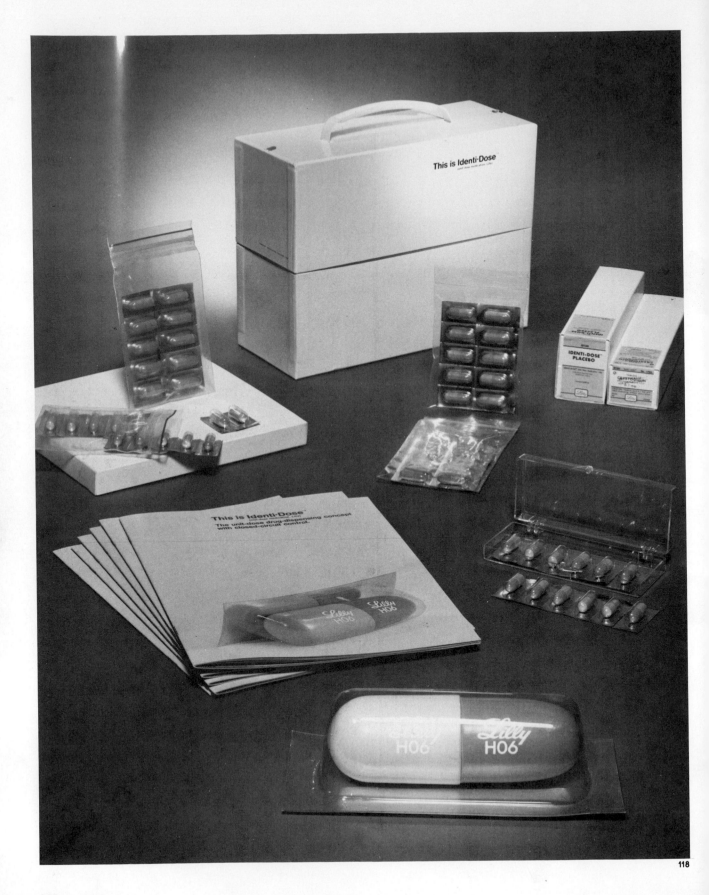

Eli Lilly & Company was a leader in the pharmaceutical industry with its introduction of the Identi-Dose® system of hospital drug identification and dispensing. Each dose is individually packaged in an air-tight, tamper-proof, plastic bubble which is backed by a paper-laminated foil bearing the label with the medicine's name. The package also serves as a container for dispensing the medication. The number on the Pulvule (HO6, in the case of Figure 116) provides double identification. The drug's exact formula can be determined by reference to a code index distributed by the company.

Another working package is the MICRO-ENCAPSULATION pocket that contains the product and also serves as an applicator. Liquid floor polishes and cleaners come in disposable urethane applicator pouches with vinyl-coated paper "hats" sealed to the urethane. The hat acts as a sponge applicator. Shoe polish is another product that is marketed in micro-encapsulation pockets. The polish is trapped between the two layers of a polyethylene/paper pouch and released upon application.

The BAG-IN-BOX is a functional and practical package that is appearing more and more frequently on supermarket

118 Redi-Vial™ allows self-contained mixing of components. Photo courtesy of Eli Lilly & Company.

119 The bag-in-box. Photo courtesy of The Great Atlantic and Pacific Tea Company, Inc.

119

shelves for many different products such as snacks, cereals, and cookies. Cartons with different types of paperboard and laminates can be used. The inner bag is made of plastic film or aluminum foil laminates.

THE RETORTABLE POUCH is the flexible package with the most universally acknowledged promise. Developed to contain a wide variety of foods and ready meals, it is now considered one of the major breakthroughs in flexible packaging.

The pouch itself presently consists of a metalized nylon and various Saran-coated films. The first user of this new packaging system was the United States Army. Further research is underway on laminations and coatings that will provide greater barrier protection from light, gas, and moisture and that will extend the product's shelf life. The replacement of foil would provide a "microwave retortable pouch." Like most sealed products, the retortable pouch will be sold in a folding carton.

The future of flexible packaging is wide open for new concepts and should present to the designer with imagination and understanding of this exciting packaging medium an exciting challenge to create new packaging formats.

designing shopping bags

The shopping bag is a walking billboard. Shopping bags are the most effective way of advertising a store, a product, or a service. They are used by banks, manufacturers, and even by political parties during election time.

There is just no limit to the imagination when it comes to designing a shopping bag. Like a billboard, it must give an instant impression. Bloomingdale's, the famous New York department store, features very unusual shopping bags, unusual in that the name of the store rarely appears on the bag. Instead, the designers created a "Bloomingdale Look," a "look" that is timely, smart, unusual, even way out. The Bloomingdale's designer knows what the new trends are in fashion, art, music, home furnishings, interior design, architecture, the latest books, plays, movies. This designer looks, searches, reads, and participates in the current scene—in other words, does design research! It takes a well-informed, imaginative designer to be able to come up with fresh new ideas.

One of the design student's favorite projects is a shopping bag. Some imaginative, student-created bags are shown on the next page.

The construction of the shopping bag is basically very

120

120, 121 Shopping bags. Photos courtesy of Bloomingdale's.

121

127

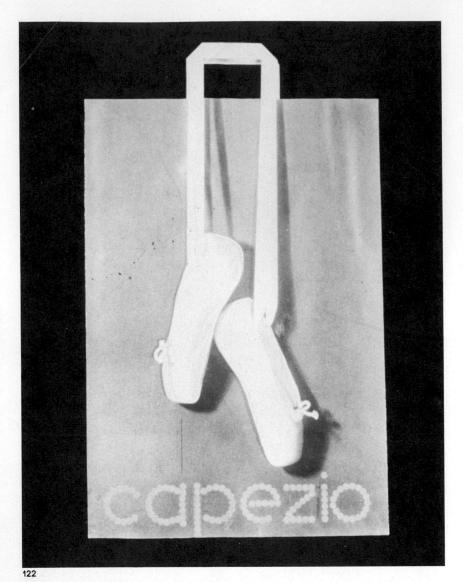

122

123

122, 123, 124, 125 Bags
created by students.

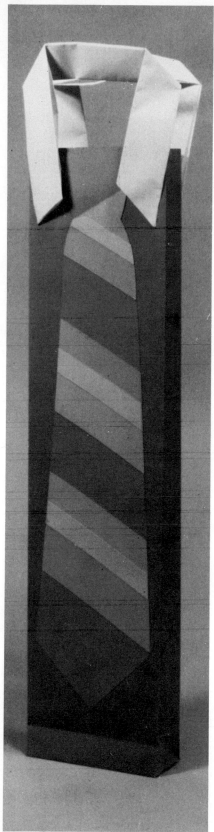

124

125

129

simple. It comes with two side gussets (pleats). Artwork can go all around the bag. The manufacturer can provide the designer with the pattern or layout, but for comps, most designers prefer to construct their own bag. TAKE ONE APART AND STUDY THE CONSTRUCTION.

To design a successful bag, it is necessary to start with a basic idea, a concept. Some thumbnails of several ideas are shown here. The student designer originated some twenty concepts in three hours in this rough, thumbnail form. Note the variety of ideas, from typography to some unusual design concepts. After the final concept is worked out, the design may be transferred to the bag through various media; photography, collage, artwork, or even Xerox reproduction are often used. There is almost no limit to the possible techniques for creating a shopping bag.

126 Bag courtesy of Gap Stores, Inc.

127 Thumbnail sketches are the first step in turning an idea into a finished design.

126

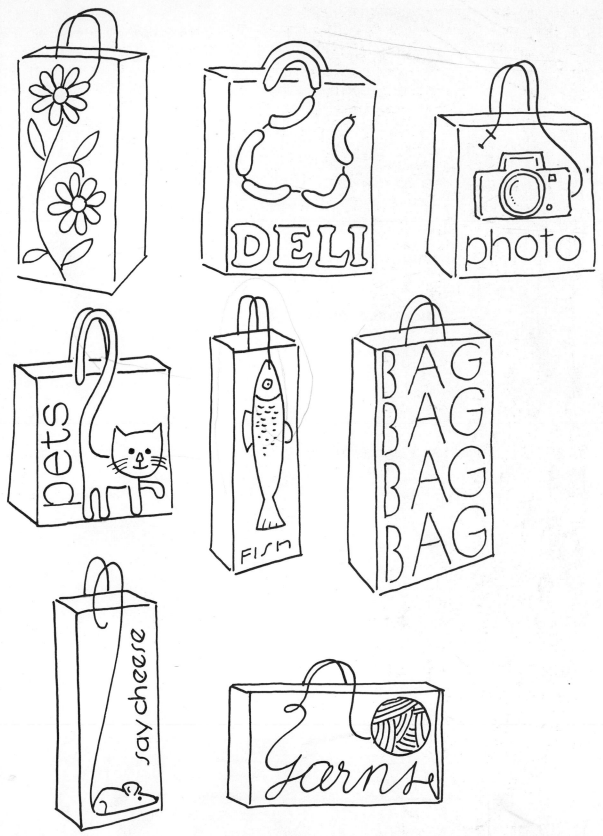

projects

Create a new concept in shopping bags. Construct a shopping bag (size optional) with side gussets. Make your own pattern. Use your imagination, your sense of design, layout, and handlettering. Create an unusual, exciting, graphically striking shopping bag!

The second project requires research and some marketing input. Originate and create a NEED for a new concept in pouch packaging. This might be a consumer product, a pharmaceutical, or even a space-flight product. Suggestions: Investigate available first-aid kits, cosmetic applicators, food dispensers. You may come up with a marketable new concept.

suggested reading

Sacharow, Stanley, and Roger C. Griffin, Jr., BASIC GUIDE TO PLASTICS IN PACKAGING. Boston: Cahners Books, 1973.

CHAPTER SIX

glass co

glass containers

Glass as an old, yet highly versatile, packaging medium. History of glass. Glass technology—past and present. Different kinds of glass containers. Stock bottles and closures. Special emphasis on the history and design of cosmetic fragrance bottles. Illustrations of some of the world's most famous and beautiful fragrance bottles. About the design procedure and model making of the glass bottle.

No other packaging medium has the versatility of glass. Glass can be opaque, translucent, clear, tinted, reusable, disposable, and recyclable. Glass is resistant to acids, most chemicals, high and low temperatures; it is thermally conductive and electrically insulating. Glass as a packaging medium is found everywhere, from practical food containers to elegant cosmetic jars and bottles. Drug and pharmaceutical products are kept pure and safe in glass bottles and vials. Soft drinks look inviting in clear glass bottles. White, green, and amber wine bottles give protection from light and preserve delicate flavors.

The popularity of glass lies in the nature of glass itself. Glass is chemically inert. It does not affect or react to the taste or odor of products packaged in it. Its smooth, non-porous surface facilitates washing and sterilizing processes and it can be sealed airtight.

Glass is made from silica (sand), soda-ash, and limestone. It is found in nature as an opaque substance called obsidian, which is produced by intense heat from lightning or, more commonly, from volcanic eruptions. The earliest known examples of man-made glass were made in Egypt sometime between 10,000 and 3,000 B.C. The first objects were beads

128

128 Soft drinks look inviting in clear glass bottles. Reproduced with permission of The Coca-Cola Company.

129

the secret of the craft. Thus began the golden age of glass. The Venetians perfected the "Cristallo" (crystal) glass, the first truly transparent glass which can be blown extremely thin and into almost any shape. By the seventeenth century, other European countries were producing free-blown glass.

In the United States, glassmaking is one of the oldest industries. In 1608, a "factory" (actually no more than a crude hut with furnace) was producing blown-glass bottles in Jamestown, Virginia. The real start of the industry came in 1739, when the German-born Caspar Wistar built a glass factory that made America's first Flint (clear) glass. Another great American glassmaker was Henry William Stiegel (Baron Stiegel). In Mannheim, Pennsylvania, he made the first and the finest art glass this country had ever seen. The famous Sandwich Glass comes from the Boston and Sandwich Glass Company, founded in 1825, which was probably one of the first glass manufacturers in the United States to produce pressed glass, the "poor man's cut glass."*

covered with green glaze. Blowing and pressing glass were unknown to the Egyptian craftsman. Melted glass was pressed into open molds, a slow, laborious process. As the craftsman's skill increased, he learned how to make the first glass containers—brilliantly colored opaque bottles and jars.

The invention of the blowpipe about 300 B.C. was to be the basis for all future glass container production methods. Glass manufacture flourished in Rome. Merchants who found that glass containers were waterproof and did not affect the taste of their contents used them to ship wines, oils, and other products on long sea voyages. During the Middle Ages, glass manufacture began on the island of Murano near Venice. As Venice grew in wealth and power, glassmaking flourished and guilds were formed to guard

*The most famous American art glass is made by the Steuben Glass Works. The company was founded in 1903 at Corning, New York, by Frederick Carder, an English glassmaker. Later, Corning Glass Works acquired the company, and it is now known as Steuben Glass, Inc. The world-famous Museum of Glass at Corning contains more than 15,000 objects made of glass.

By the early nineteenth century, production had advanced to using molds in conjunction with blowing to form containers of different sizes and shapes. A "gob" was lowered into the mold and blown to conform to the mold contour. Somewhat later, the glassmakers began to engrave the molds to create designs, shapes, and forms. Between 1820 and 1870, manufacturers produced numerous "historical" or "memorial" flasks, engraved in the likeness of famous people; later on, emblems and mottos became popular as well. Special political and social events were commemorated in glass containers that were originally made to hold liquor and later kept as ornaments. Engraved bottles were functional as well as decorative, providing information as to manufacturer, contents, and directions for use. Most were used for patent medicines and liquor.

131 Mason jars have been in use since 1858.

132 Plastic closures. Courtesy of Monsanto and W. Braun Company.

In 1858, John Landis Mason patented the famous Mason Jar. By 1880, commercial food packers began to use glass containers. Milk was moving from the metal pail into elegant glass bottles, which were designed by Dr. Thatcher of Potsdam, New York. In 1899, Michael Owens of Toledo, Ohio, invented the first semi-automatic bottle-making machine, the most significant advance in glassmaking since the discovery of the blowpipe. By 1904, the Owens bottle-making machine was completely automatic. It produced containers of equal length, weight, and capacity, which made uniform glass production a reality.

As the glass containers gained in popularity, closures had to be developed for specific containers and products. At first, corks were the most widely used closures. Wine bottles were molded with a ring on the neck to secure the cork with string or wire. Most patent medicines, cosmetics, and alcoholic beverages were "corked." The "Lightning Fastener," a glass lid fastened by snapping wire bale over a groove in the lid, was very popular. Metal screw tops, introduced on the Mason Jar, were quickly adapted to other containers and bottles. With the invention of the rubber sealing ring that guaranteed an airtight closure, other sealing materials

131

like cork, cardboard, and papers were adopted.

With the introduction of carbonated soda water, the forerunner of bottled soft drinks, it became necessary to invent a closure that could withstand the pressure produced by its contents. William Painter's invention of the metal cap with a thin cork liner proved to be airtight. Called the "Crown Cap," this was a simple, shallow metal disk with a flared, fluted "skirt" that was forced against the ring on the bottle neck for a tight seal. It could be easily removed by prying it off with a bottle opener (another new invention). Lug caps, vacuum caps, and friction-fit caps of plastic or metal are all outgrowths of the development of closures for containers. Plastics added a new dimension to closure development. Thermoset plastics, ureas, and phenolics were used in the early 1900s. They were favored for rigidity and for chemical resistance, making them suitable for use with many chemical and pharmaceutical products. Thermoplastic closures are less expensive than thermosets and can be recycled. Materials include polyethylene, polystyrene, polypropylene, and SN and ABS copolymers. Thermoplastic closures can be designed with dispensers like flip-tops, spouts, and other pouring devices.

production processes

Four basic processes are used in producing glass: blowing, drawing, pressing, and casting.

blowing uses compressed air to form molten glass in the cavity of a mold. Most commercial glass bottles are produced on completely automatic equipment by this method.

drawing is a process in which molten glass is pulled through dies or rollers that shape the soft glass. Tubes and sheet glass are produced this way.

pressing is used to press the molten glass against the sides of a mold.

casting uses gravity or centrifugal force to cause a gob of molten glass to form in the cavity of a mold.

133

The portion of the bottle that contains the opening and accommodates the closure is called the FINISH. This word is a standard term in the glass container field. There are thread, lug, and friction finishes. There are special types of finishes: sprinkler tops, roll-ons, pour-outs, and snap caps. In the interest of economy and efficiency, glass finishes have been standardized. The standard finishes are designated by number. For example, GCMI finish 400 is a shallow, continuous thread (the kind most widely used). These standards are on file with the Glass Container Manufacturers Institute (GCMI) of Washington, D.C. Glass finish sizes are designed by the size, in millimeters, of the outer diameter of the bottle.

Structurally, glass containers and bottles have changed significantly in the last 30 years. Most glass containers have shed about 55 percent of their weight. Larger containers have become feasible through design and material improvements and by plastic foam coatings that add an element of safety in the event of breakage.

bottles are the most extensively used glass containers. Shapes can be cylindrical, oblong, or rectangular. The neck is almost always round to make pouring easy and to lend itself to effective closure. Today, beverages such as wine, beer, liquor, and soft drinks account for well over half the glass containers produced in the United States. The demand for non-returnable (NR) bottles for beverages began to grow during the 1940s. The development of the resealable "twist-off" closure encouraged the popularity of the large-size NR bottle.

jars are similar to wide-mouth bottles, and certain types and shapes lend themselves to various products, mostly cosmetics and food. The large opening accommodates utensils and fingers. Jars usually have a low center of gravity and lend themselves to easy handling.

tumblers are related to jars, shaped like a drinking glass, and used in packaging food (mostly

133 Thirty-two-ounce glass bottle for Coca-Cola with Plast-Shield coating. Reproduced with permission of The Coca-Cola Company.

134 Jars are similar to wide-mouthed bottles. Photo courtesy of W. Braun Company.

135 The large openings in jars accommodate utensils and fingers. Photo courtesy of Tang™ Orange Flavor Instant Breakfast.

134

135

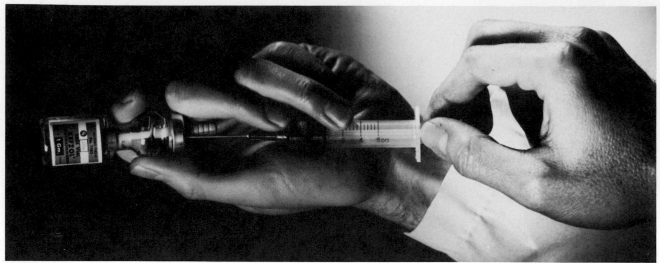

136

items like jams, jellies, spreads, and sauces).

jugs are large bottles with short necks and carrying handles. The most common capacities are the gallon and half-gallon sizes.

carboys are industrial, heavy-duty shipping containers (mostly for chemicals), usually of three- to thirteen-gallon capacity. They come with a wooden crate holder.

vials are small glass containers. They are flat-bottomed, tubular in shape, and have a variety of neck finishes. They are widely used for antibiotics and other pharmaceuticals. Plastic closures or other tamper-proof devices are used on these exceptionally strong containers.

Eli Lilly & Company's Redi-Vial ™ (Figure 136), a vial with separate compartments for dry medication and sterile diluent, allows self-contained mixing of these components. Pressing the plunger-cap of Redi-Vial dislodges a rubber stopper and allows the premeasured drug and diluent to be mixed. The mixture then may be drawn into a syringe, ready for use. Redi-Vial improves accuracy because the drug and diluent are premeasured, reduces the chance of contamination, and saves time because it is not necessary to transfer diluent from one vial to another.

Because of the nature of modern drugs, drug packaging is becoming more complex. All types of products—pills, powders, ointments, liquids—are packaged in glass. Each area of drug packaging has its different glass container requirements.

ampoules, made from glass tubing, are used for serums and injectable drugs. After the product is poured into the ampoule, the open end is melted and sealed shut. The ampoule must be broken at a designated breakline before the contents can be used.

Decorated glass containers and closures can be attractive as well as functional. Bottle labels can be front, back, full wrap, or neck bands. Pressure-sensitive labels are also used on glass, especially on cosmetic containers. Silk-screening in ceramic and metallic colors is another frequently used method for decorating glass. Other special effects such as frosts, decals, and transparent colors are possible. The latest innovation in glass decorating is the use of thermoset powders, which can be cured at relatively low temperatures to provide an overall finish in an unlimited range of colors. Closures are equally versatile and can be printed, screened, embossed, hot stamped, or vacuum metalized.

The elegance and beauty of glass make it ideal for packaging cosmetics. Cosmetics, like drugs or pharmaceuticals, require chemically inert containers for long shelf life. The cosmetic industry packages products having a broad range of prices. The cosmetic bottle or container reflects many requirements. A fine fragrance or lotion is an exclusive, luxurious object, produced with high regard for quality and excellence, and the bottle, jar, or aerosol is the visual symbol of the product. The consumer feels that the product expresses his or her personality, and its container is proudly displayed in the home.

136 Identi-Dose® system of hospital drug identification and dispensing. Photo courtesy of Eli Lilly & Company.

137 Cosmetic packages are exclusive and luxurious objects produced with high regard for quality and excellence. Photo courtesy of Parfums Hermès.

137

designing a fragrance bottle

There are two basic approaches to the visual design of any bottle or container: choice of the right stock bottle with the appropriate closure, or creation of an original (private-mold) concept.

Each procedure has its merits and advantages. The stock bottle is a wise choice when cost and time are the major factors. There is a wide selection of stock bottles and closures available at standard prices, and the competent package designer recognizes that the appropriate stock bottle can give him wide scope for imaginative design. The term "stock bottle" does not necessarily mean that these items are readily or immediately available. Some stock molds must be run in minimum manufacturing quantities. Stock bottles can be run in flint, opal, or even in color.

The combination of the right stock bottle and closure with a distinctive label or decor can result in an excellent presentation. Another advantage of the stock bottle is that the glass manufacturer or distributor can supply the designer with actual samples on which to create a wide variety of designs, instead of having to make sketches, mock-ups, or expensive three-dimensional models. Some stock houses publish profusely illustrated, encyclopedic catalogs, with more than a thousand items to help the designer create a better package. Basically, the ideal stock bottle is simple, well-designed, and pleasing in proportion and contour, unlike the private-mold, custom-designed bottle, which can and should be distinctive and unique.

The basic approach to the design of a private-mold bottle is that the fragrance's name is incorporated in the design theme, and that the fragrance and the bottle are treated as equal partners, a work of art in sophistication and quality. Exclusiveness should be expressed by select decoration techniques and a feeling of richness. Elegance is induced by refinement, sophistication, and good taste. The creative

138

designer aims at a classic, timeless creation and avoids amusing, clever, or cheap novelties. A good name for a fragrance can largely determine its success. If the name can also be interpreted in the design of the bottle, it will greatly contribute to the success of the product itself. The bottle will become the visual symbol of the product, easily recognized and identified.

Shown here are some of the world's most famous fragrance bottles, ones that through the years have become classics.

There are fashions in bottle design as there are in dress, from refined simplicity to the most complex shapes and combinations. In recent years, the long, exaggerated closure over the bottle has been the most popular. Years ago, round and flared closures were fashionable. To be able to

139

140

138 A selection of stock bottles. Photo courtesy of W. Braun Company.

139 Photo courtesy of Chanel, Inc.

140 Photo courtesy of Elizabeth Arden, Inc.

147

147 A lucite model is an exact replica of the proposed bottle and is prepared by an expert model maker.

take several weeks, or even months, to prepare the proper mold that will produce this most original, unique bottle, the ultimate result of YOUR design and the teamwork of many packaging experts.

suggested reading

Books:
Helen and George McKearin, TWO HUNDRED YEARS OF AMERICAN BLOWN GLASS. Crown Publishers: New York, N.Y., 1946.

Trade Publications:
GLASS PACKAGING, PACKAGE ENGINEERING

for further reference

Trade Associations:
Glass Packaging Institute, Washington, D.C.
Society of Glass Decorators, Port Jefferson, N.Y.

Stock Bottles and Closures:
W. Brown & Co., New York, N.Y.

Custom Bottles:
Wheaton Industries, Millville, N.J.
Carr-Lowery Glass Co. (Div. of Anchor Hocking), Baltimore, Md.

Collections:
Metropolitan Museum of Art, New York, N.Y.
Corning Glass Works, Corning, N.Y.

CHAPTER SEVEN

cans, tubes, and aerosols

An exciting chapter about one of the oldest and youngest packaging mediums. The history of metal cans and can making. How to design a metal can. Brand name as package identification. Various printing and labeling techniques. What you should know about metal and plastic tubes. How to apply principles of good graphics to tubes. An important (and sometimes controversial) packaging medium: the aerosol. The advantages of the aerosol package. New, efficient, and safe propellants.

Napoleon could be called the godfather of the metal can. In 1795, he authorized a competition awarding 12,000 francs to anyone who could devise an effective way to preserve food. Nicolas Appert developed a method of putting partially cooked food in glass bottles with cork closures and then immersing the bottles in boiling water. Appert won the prize and published a paper on the theory of the evacuation of air and use of heat to keep and preserve food.

In 1810 an Englishman, Peter Durand, received a patent for putting foods in "vessels of tin, glass, pottery and other metals."* Since glass is fragile, Durand proposed a cylindrical tin canister made of iron with tinplate coating.

The first commercial cannery was set up by two other Englishmen, Bryan Donkin and John Hall, in 1813. They began to supply the British Army and Navy (which eventually defeated Napoleon at Waterloo). The early metal can was handmade, the side seam and end sections soldered. Food was inserted through a hole on the top and a small disk was then soldered over the hole. The disk had a tiny hole to permit the escape of air which had heated and expanded during the filling process.

*Crown patent, 1810, England.

Soldering this hole completed the canning process.

By 1819, the can had crossed the Atlantic. New Yorker Thomas Kensett and Bostonian William Lyman Underwood were already canning in glass. But it was Kensett's 1825 patent on the can that started the development of the "tin can." By 1847, the flanged ends were mechanically formed. The Civil War helped to further advance.

148

148 United States Army "C" rations.

149 Some early cans, circa 1920. Container Corporation of America, Design & Market Research Laboratory, Packaging Museum, 400 East North Avenue, Carol Stream, Illinois 60187.

149

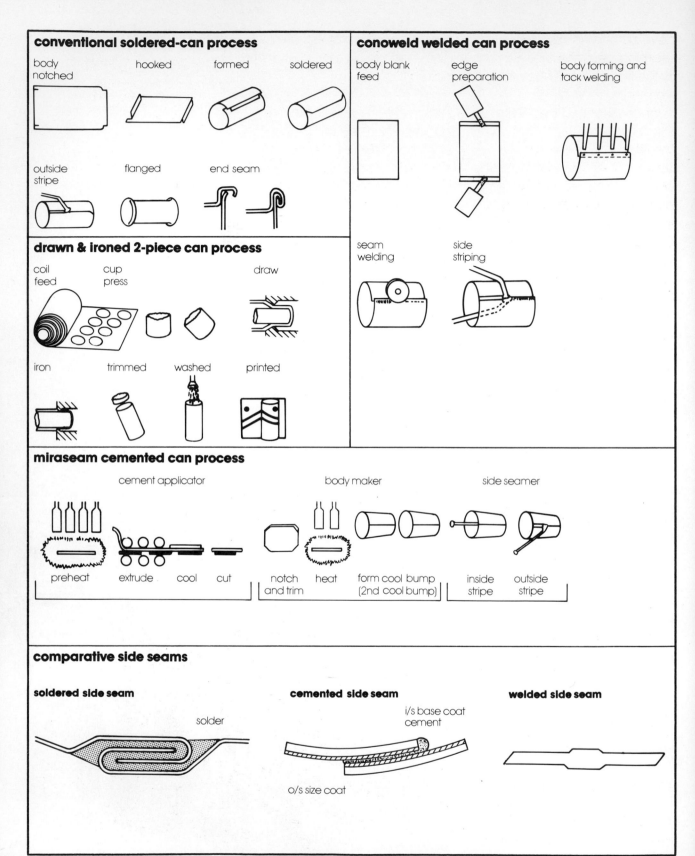

conventional soldered-can process

body notched hooked formed soldered

outside stripe · flanged · end seam

drawn & ironed 2-piece can process

coil feed · cup press · draw

iron · trimmed · washed · printed

conoweld welded can process

body blank feed · edge preparation · body forming and tack welding

seam welding · side striping

miraseam cemented can process

cement applicator · body maker · side seamer

preheat · extrude · cool · cut · notch and trim · heat · form cool bump (2nd cool bump) · inside stripe · outside stripe

comparative side seams

soldered side seam

solder

cemented side seam

i/s base coat cement

o/s size coat

welded side seam

development. In 1867, side seaming became mechanical. By the turn of the century there were more than 1,800 canneries in the United States.

The most important development of the early 1900s was the open-top cylindrical can, which would be the standard for years to come. Over the years oval, oblong, square, rectangular, and fluted cans would be made for products ranging from paints, tobacco, and industrial products to spray cans.

The metal can is not what it once was. Today's metal cans consist of tin-free steel lined inside with various plastic resins (acrylics, vinyl, epoxy, alkyds, and others). The introduction of tin-free steel presents an entirely new concept in can making. The side seam is held together by organic cement or welding, rather than the conventional soldering method. The absence of solder permits the decoration of the entire can body.

With its light weight and resistance to corrosion, aluminum is used to package beer and carbonated soft drinks. The two-piece drawn aluminum containers have no side seam and can be printed around the entire can body. Although steel is by far the predominant material used for cans, a variety of aluminum

fiber (paper) containers are also used, usually for juices and baked foods.

In almost all types, the metal can incorporates an easy-open device. Ring-pulls and pull-tabs are typical opener devices. Despite the ever increasing number of easy-open devices, the old, familiar can opener is not a complete casualty, because most metal cans, especially those containing foods, are still opened by this "old" method.

In the booming beverage industry, the use of metal cans grew to a record 84 billion units in 1976! The future seems almost unlimited for metal cans.

150 Can-making processes.

151 Examples of contemporary cans. Reproduced with permission of The Coca-Cola Company and The Great Atlantic and Pacific Tea Company, Inc.

151

designing metal cans

The development of new printing techniques is adding sales appeal to metal containers. It is important for the designer to be aware of these new developments before planning the graphics for the containers.

For years, the paper label was a very important part of the soldered metal can, and a large percentage of cylindrical metal cans, especially food cans, still use paper labels. Modern, high-speed printing processes permit lithographic printing of graphic illustrations or photographs in as many as six colors. Current innovations include embossed containers. The raised surface adds a sparkle to the metal container that is impossible to achieve with a paper label. The use of contrasting, transparent colors certainly adds to the appearance and to the graphic impact of the shiny metal container.

The most important aspect of the design project is the identification of the product by its BRAND NAME. Because the can is a round object, visibility and readability become a problem. A flat design appears totally different to the eye than when it is wrapped around a round can. This is basically an optical problem and it must be treated and solved as such.

Package identification by brand name is a study by itself. The brand name is a verbal part of the trademark, and can be symbolic or pictorial. The brand name on the round package can be successful only when it is appealing to the eye and visually gives an expectancy of value and quality. Many of the popular soft drink cans are designed with this important principle in mind. Coca-Cola, 7Up, Pepsi, and Sunkist are especially striking designs.

The illustration or the photo on the label must perform the function of a "transparent" container—the consumer must "see" what's in the can. Full color photography or illustration can serve the designer well. The designer must be a virtuoso, using good type or effective handlettering for the brand name in combination with striking product illustration. (Take another trip to the supermarket and see some excellent examples in the canned food aisle.)

To design a successful can, you must know your product. Imagine you are going to design a soft drink product called COOL. This product comes in an aluminum can. As the design will be printed directly on the can, you can take advantage of the shiny, reflective surface and print with transparent colors.

First, you must develop the logo (COOL). It sounds simple, but there is an optical problem of lettering on a round can that has to be solved first. Start your design concept with an actual can. Wrap a paper around the can, set the can up on a shelf or table, and step back and look at the can. Determine the area your eyes can see without turning the can. This should give you an idea of how to work your logo, how much space you need, whether you should try a vertical or slanted logo, whether you should use symbols, illustration, or some other design device to give your product "shelf-appeal." All this you will have to work out in a rough form, testing and trying your concept directly ON THE CAN.

Now, you may have three or four good design possibilities. Render these on paper, foil, or acetate in transparent colors and wrap them around the can. These are your comps. Present your comps to the client. (Always keep the unused roughs—just in case.)

Once the design is chosen, the next step is a simple mechanical, usually specified by the client's production department.

The important thing to remember is that clever artwork or a wild color combination are no substitutes for an effective, smart logo. Spend time and effort to develop YOUR logo. Clients and art directors are wary of store-bought type. Your understanding of typography and your lettering skill are your most important assets when you design logos, trademarks, and brand names for the product in the can. Design a stunning COOL can!

collapsible metal tubes

The first collapsible metal tube was developed in 1841 by an obscure painter for dispensing artist's colors. In the 1890s an American dentist, Worthington Sheffield, invented a new dental cleaner—toothpaste. Sheffield packaged his new product in a collapsible metal tube. Today, more than half of the metal tubes manufactured are used to dispense toothpastes. Tubes are lightweight and provide excellent product protection and easy disposability. Tubes are safe, sanitary dispensers for pharmaceutical and cosmetic products.

The most important consideration for metal tube design is product-package compatibility; the type of metal to be used and the lining (if it is necessary) must be appropriate for the product they contain. These are technical and chemical decisions, specified by chemists and engineers. There are three basic tube materials:

aluminum tubes are used for toothpaste, shaving cream, toiletries, and food.

tin tubes are real tin—strong, durable, and chemically inert. They are used when high demands of compatibility must be met, especially for medicinal-pharmaceutical ointments.

lead tubes are less expensive than aluminum or tin. These tubes can be lined with various coatings, depending on the substance they will contain. Paints, adhesives, and chemicals are some of the products that come in lined lead tubes.

Tubes are produced from stamped or punched metal slugs that are then fed into extrusion presses. The tubes are then conveyed to a machine that TRIMS THEM FOR CRIMPING AND THREADS THE NECK ENDS for the appropriate closure. The tubes are now conveyed to an offset printing machine that can print up to four colors. After drying, the tubes are automatically capped.

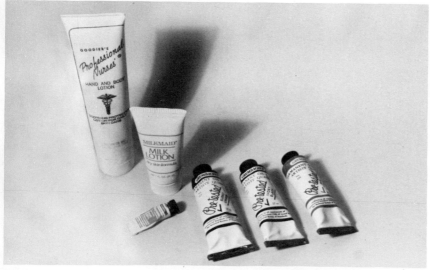

152

plastic tubes

The main features of plastic tubes are that they are unbreakable, leakproof, transparent or opaque, and inexpensive. Plastic tubes are manufactured of extruded polyethylene, polypropylene, or a combination of plastics.

Most plastic tubes come in stock sizes and shapes, but color and transparency or opacity can be specified. Unlike metal tubes, plastic tubes can utilize a great variety of decorative processes: offset printing, silk-screening, hot stamping, heat transfer, and embossing. Plastic tubes can be lined internally.

Water-based and oil-water emulsions like shampoos, lotions, and hair-grooming products can be packed in plastic tubes. Filling and handling methods are similar to those for metal tubes; however, plastic tubes are sealed by heat or ultrasonics.

Plastic tube design can be versatile and requires an imaginative, creative approach. One recent development in tube utilization is a viscous candy that is eaten directly out of the tube. Cake decorating tubes are another creative effort to utilize plastic tubes.

Skill in lettering and typography can help the designer create a truly unusual and attractive package. The toothpaste tube itself is an excellent example of the proper use of type and lettering. Major toothpaste manufacturers like Colgate, Aim, and Gleem periodically redesign their tube, updating the logo and the overall look of the tube.

In the fields of personal products, hardwares, and household products, plastic tubes are becoming more and more important. They are low-cost, versatile, and practical applicators of products.

To prepare mechanicals, the designer should be provided with the proper layout by the tube manufacturer. For comps, the manufacturer will provide the designer with blank tubes and closures.

aerosols

The modern aerosol is the result of research in the early 1940s by two Americans, Dr. Lyle D. Goodhue and William N. Sullivan. A metal container was pressurized with gas and fitted with a pushbutton dispenser to deliver a spray of insecticide. This invention, called the "Bug Bomb," was used during World War II in the Pacific.

After the war, the aerosol was converted to peacetime applications: hair sprays, fragrances, toiletries, food, medicines, paints, and many other consumer and industrial products. The new system was called the pressure-packaging system.

An aerosol consists of an airtight, valved container (metal or glass) filled with a formulation of a propellant (gas) and the active ingredient (the product). When the valve is operated, the gas pressure in the package pushes the propellant and the active ingredient through the small pinhole opening in the valve. Fine mists, sprays, puffs, foams, or dust can be produced, depending on product, propellant, and the type of valve.

There are three types of propellants: fluorocarbons, hydrocarbons, and compressed gasses. The most efficient propellants are the fluorocarbons. Unfortunately, recent scientific findings concerning the interaction between fluorocarbon propellants and the ozone barrier in the ionosphere have prompted the Environmental Protection Agency (EPA) and the Food and Drug Administration (FDA) to discourage or ban the use of fluorocarbons. The result will be substantial reduction in the use of aerosols. Hydrocarbons (propane, butane) and compressed gases (carbon dioxide or nitrogen) may be able to take over some applications. Research is going on to find other efficient and safe propellants. The aerosol is a major advantage for drugs and pharmaceuticals: because the product is sealed in the container until used, contamination from the air is eliminated and full strength of the formulation is preserved.

The temporary setback of the aerosols has resulted in

153

154

important developments in mechanical spraying pumps. The pumps are designed to produce aerosol-type mists or sprays. This pressure packaging uses glass or plastic containers, which has reduced the major role of metal cans in the aerosol field. There are significant developments in various piston and bladder devices that may use metal cans with compressed air.

suggested reading

PACKAGE ENGINEERING, 1978, 1979.

MODERN PACKAGING, 1978, 1979.

projects

Design a type of plastic tube that can stand on the closure.

Develop a tube that has an applicator closure.

Originate a product (household or cosmetic) that will utilize a tube.

Keep in mind that design and typography should be distinctive even when the tube is upside-down, partially rolled, or empty.

155

156

CHAPTER EIGHT

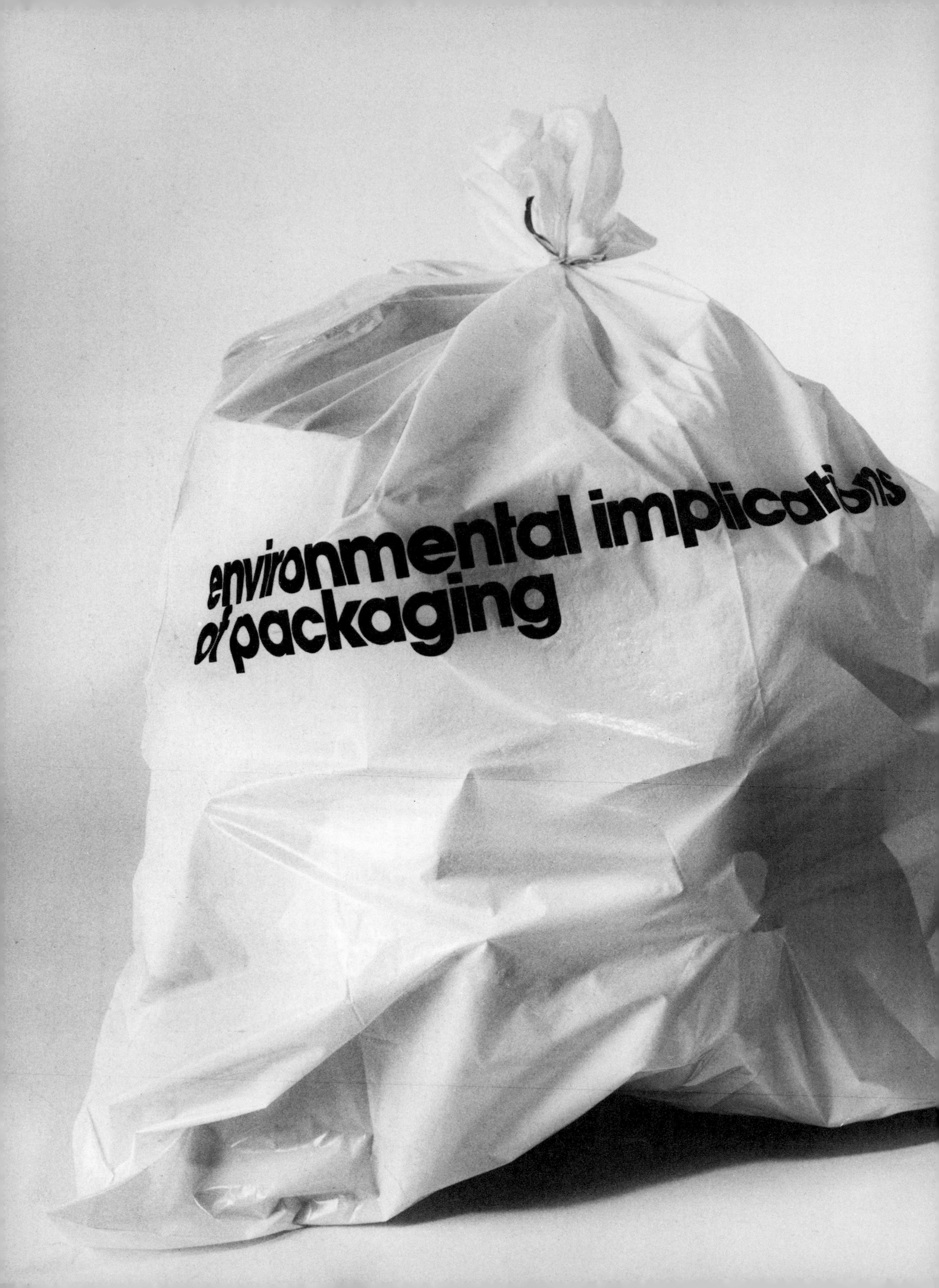
environmental implications
of packaging

environmental implications of packaging

A serious study of packaging and the
environment. The basic federal laws on
packaging and packaging materials.
Introduction to the problems of solid waste
disposal, recycling, and resource recovery
programs, with charts of government and
industry estimates. The designer's role in
creating ecologically acceptable packages.
Design projects to create reusable packages. A
list of federal agencies, books, and trade
periodicals to follow for the latest news on
federal laws and regulations.

Our early ancestors tossed bones, broken pottery, and other useless rubbish into heaps near their caves. When the heaps of refuse grew too large, the people just moved on.

In ancient Rome, Pliny complained that "the air is foul." In medieval times, streets were strewn with garbage, although there were strict laws against littering.

Today, we are still basically a throwaway society. We produce wastes in ever-increasing volumes. Only recently have we become aware of our decreasing forest lands and our polluted air and water. With this awareness, legislation dealing with solid waste disposal has been introduced and packaging material disposal has become a vital issue.

Present municipal refuse amounts to about 145 million tons annually (1978) and is growing at about 3 percent per year, according to Environmental Protection Agency estimates. About 35 percent of this municipal waste is packaging materials. To coordinate the assembly of solid waste studies, the packaging industry has established the National Center for Resource Recovery, Inc., in Washington, D.C. Individual industries have established pilot projects for the separation of waste materials, for recycling, and for creating energy from waste sources. The federal government introduced the Solid Waste Disposal Act of 1965 and the Resource Recovery Act of 1970, which have made funds available for local pilot projects. The Resource Conservation and Recovery Act of 1976 mandates national action on recovered materials. Citizens' groups have established voluntary collection centers for newspapers, metal cans, cardboard, and glass bottles in communities across the United States.

understanding solid waste disposal

Disposal and recycling, like so many modern problems, are difficult to solve because of their size and complexity. Landfills, "the old town dump"—like those of Stone Age man—do become filled up eventually, and areas for new fills are difficult to find.

The search for better methods of refuse disposal usually begins with burning. Incineration reduces the volume, but causes pollution in the atmosphere. Burning in open dumps is prohibited and municipal incinerators have been fitted with pollution control devices. The combustion of refuse creates valuable heat energy. It can heat buildings, turn turbines, and generate electricity. Some materials also can be recycled and processed before incineration.

In Europe, where there is even greater scarcity of space for landfills, there are plants that reduce the volume of refuse while extracting energy from it. A plant in Berne, Switzerland, can burn up to 200 metric tons of refuse a day while producing steam, hot water, and enough electricity to run a hospital, several schools, a large apartment complex, a railroad station, and a chocolate factory. Paris, Frankfurt, Amsterdam and most other major European cities (as well as many smaller ones) now have similar plants. The concept also has been adopted in Japan, Australia, and Canada. In the United States, adoption of refuse-energy systems is increasing. A recent application in the United States is typified by the 1,200-ton-per-day installation in Saugus, Massachusetts, which delivers steam for electric generation and plant use in a nearby industrial facility.

the recycling story

Because packaging represents about 35 percent of municipal refuse, resource recovery of essential materials is an important task. About 2 million tons of the 9 million tons of newsprint consumed annually are now recycled. Used corrugated boxes and grocery bags can be used to make new corrugated boxes. All paperboard can be reused to make new folding cartons.

Diverse packaging materials are difficult to sort and separate, but BOTH COMMUNITIES AND INDUSTRIAL SOURCES ARE CAPABLE OF COLLECTING AND REUSING ALL MATERIALS with the equipment and systems that are today available.

Recycling of aluminum by communities accounts for a 12 percent recovery of all discarded aluminum cans. Six percent of all steel cans are recycled, and 3 percent of all glass is recovered at volunteer collection centers. (Glass must be separated by color and must be free of metal caps and other non-glass materials.) The picture is bleaker where plastics are concerned. Only the plastics industry can effectively recycle its own surplus and waste materials, especially the thermoformed plastics. Since both plastics and glass are NOT biodegradable, and since the raw materials (especially glass) are not too costly, proper incineration is the answer for both economic and environmental reasons. Although incinerators now in use are somewhat antiquated, proper equipment is available for incineration without pollution. The issue appears to be political and economic—lots of talk and little action. Waste CAN be converted into energy. And there's money to be made in resource recovery!

A long-term outlook for more industrialized systems for large-scale recovery projects could be a lucrative, profitable business. It has been estimated that the potential revenue from municipal waste is about $1.5 billion annually, and the total cost of a nationwide network recovery system has been estimated at $3.8 billion. Such an efficient disposal and recovery system would give us

a cleaner environment, permit the recovery of valuable materials, and give us a much-needed additional source of energy. It is up to communities and industry to keep our land clean and free of pollution. The technology is available for just this, and it is a profitable, long-term investment.

RECOVERY OF RESOURCES BY RECYCLING IN THE PACKAGING INDUSTRY			
PRODUCT	GROSS DISCARDS (1,000 TONS)	RECYCLED (1,000 TONS)	PERCENT RECYCLED
Glass	12,755	380	3
Steel	5,420	300	6
Aluminum	900	105	12
Paper, Board	26,625	5,095	19
Plastics	3,075	0	0

SOURCES: Environmental Protection Agency, Office of Solid Waste, Resource Recovery Division. Franklin Associates Ltd. Government and industrial estimates.

suggested reading

"Refuse-Energy Systems with Resource Recovery as Alternatives to Landfill." New York, N.Y.: White, Weld & Co., 1977.

new options

157 Printed corrugated boxes can be reversed and made into playthings.

It is the responsibility of the designer to utilize available new materials creatively and flexibly as well as profitably. Waste is not an economic blessing; it is an economic disease that sooner or later will result in economic and ecological disaster. The designer must be aware of our resources and our developing technology and design with dedication and good taste. Some of the best examples of this kind of thinking are the small, compact, gas-efficient cars; reusable packaging; plastic containers designed for future storage of various small or large objects; and commercial glassware that becomes an attractive houseware or decorative object. The all-purpose plastic bag is one of the most practical reusable packages. But there are still NOT ENOUGH practical reusable packages. The designer should investigate new possibilities.

Industry claims that the consumer WANTS the convenience of disposable packaging. The non-returnable beverage containers were introduced to satisfy public demand, but they will no longer satisfy an ecology-conscious public and its legislators. Some milk distributors in suburban and country areas are once again utilizing the conventional glass container. While this method may be profitable in geographic areas where the easy collection of these containers is possible, recycling and recovery present a more complex problem in our urban areas. Disposal, recycling, and recovery systems will become a reality only when industry and the communities are able to work together to solve the problem through legislation, technology, and consumer cooperation.

The designer will play a major part in the coming change of packaging systems. In the past, there have been many attempts by designers to create attractive reusable packages. One of the best examples was the packaging of liquid detergents in historical glass bottles. The candy and confectionary and gourmet-food industries package their products in attractive, reusable accessories (dishes, baskets, toys). The cosmetics industry uses beautiful, sophisticated gift packaging, especially during the holiday season, often with reusable containers. These are "natural" packaging devices, available from the gift industries or imported from overseas. The designer should develop new, practical, profitable ideas with the new materials.

Here are some ideas. Why can't someone develop an inexpensive method of printing the INSIDE of a corrugated box

157

problems
and
solutions

problems
and
solutions

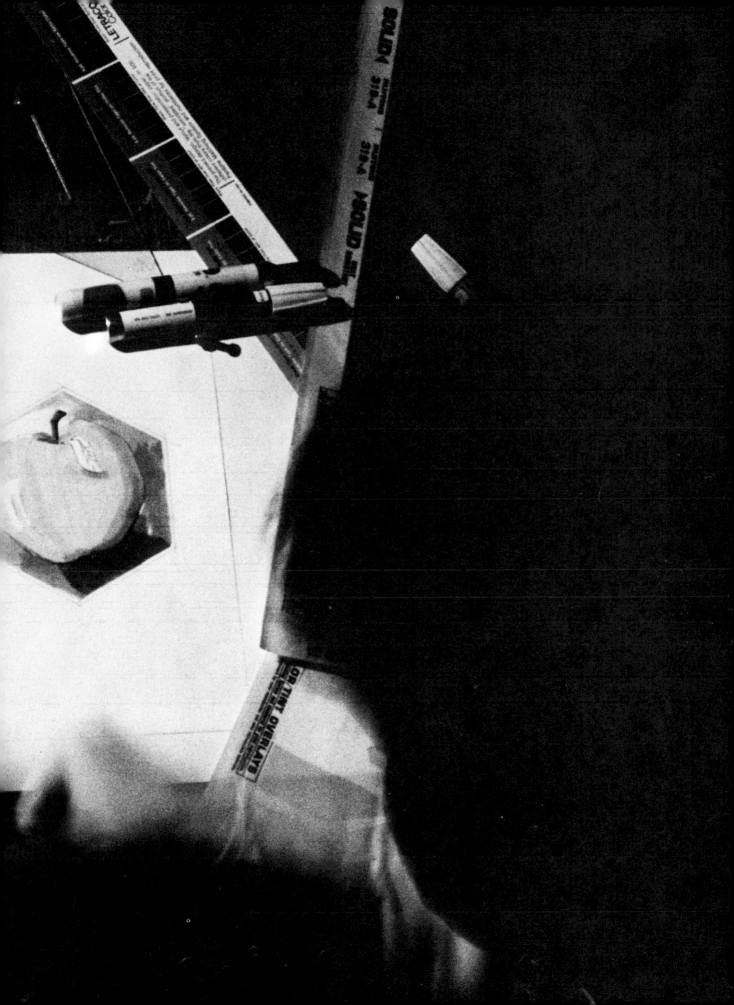

problems and solutions

Brief descriptions of fourteen real and unusual
package and product design problems and
how they were solved. Step-by-step descriptions.
Illustrated with photographs.

with paper and board

problem 1

Create an inexpensive sales promotion and display item for package stores, supermarkets, and drugstores to promote a variety of items (beverages, food, toiletries, etc.). A mobile type of display would be considered. Limited budget requires careful planning: minimal assembly, low-budget printing costs. Some type of simple die-cut device would be preferable.

SOLUTION. A simple, die-cut, printed item based on the "center of gravity" theory. A variety of amusing and attention-getting items can be made from preprinted, die-cut cardboard (40-60 point): birds, acrobats, small animals. These die-cut figures are placed on a pointed object (dowel, wire, or top of a bottle). When they are properly balanced they will move in the slightest breeze. These are inexpensive to produce, with virtually no assembly required. In addition to their display value, these items can also be reproduced on the backs of food cartons as a premium for a children's toy.

How does one find the center of gravity? The ordinary scale is the best example of this theory. The balancing point of any object IS the center of gravity. Study these die-cut objects carefully, especially the birds. You should experiment and make your own devices. Once the item is balanced it will move by itself. Space for advertising copy can be added. This item creates a very strong visual impact.

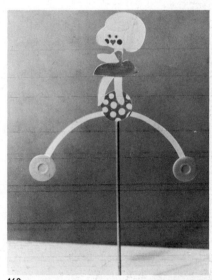

160

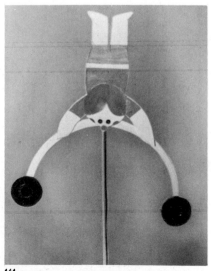

161

162

163

problem 2

Design a household product and an activity toy for small children made from commercial corrugated board.

SOLUTION: Large areas in an apartment or house can be divided into temporary smaller areas to create a sense of privacy. Corrugated board (A. B. or C flutes) can be converted into room dividers and decorative screens. Original art or posters can be used effectively to create decorative designs of a great variety. Since corrugated board is a lightweight material, it is recommended that some weight be utilized at the base of the screen. A 2"×1" wooden strip glued to the bottom would help to stabilize the screen. For a more permanent look, "framing" of the screen would be helpful.

Large 24"×24" squares of B-flute corrugated board, slotted and decorated with colorful designs, can be used as a giant building toy. A minimum of six to eight squares is required to create twelve different shapes (including a playhouse). Corrugated boards in sheet form are available at your local corrugated box dealer.

164

problem 3

Create a carton for a doll (or small toy) that converts into a plaything.

SOLUTION: A conventional tray-type carton (about 20-point board) with some accordion-folded screens on both sides of the carton, folded into the carton as a lid, and heat-sealed for a sales package. When the heat-sealed package is opened, the screen can be folded out to create an "environment" for the toy. For all practical purposes, the entire package can be made in one piece.

165

166

problem 4

Create simple die-cut greeting cards and notepapers using die-cutting for special effects. Minimum of printing. One color preferred.

SOLUTION: Simple die-cut shapes and figures on the front panel of the cards can be designed to create a three-dimensional effect when the card is opened. The inner panel is printed in one color. The die-cut piece from the front panel can be used as an additional design element when adhered to the inner panel. A variety of textures and foil and colored board can be used.

167

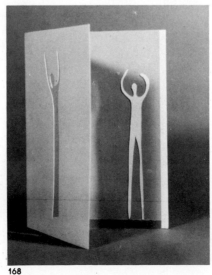

168

169

problem 5

Create a series of novelty folding cartons that will contain a variety of child-oriented merchandise (candies, food, vitamins, etc.).

SOLUTION 1. Nested cartons. These are ordinary folding cartons with open bottoms to nest in each other. The last and the smallest one contains the merchandise (in this case a small bottle of vitamins).

SOLUTION 2. A puzzle carton. A series of 3″ × 3″ × 3″ cubes. Each carton tells a story in comic-strip form. Several cartons together make an impressive puzzle. Printed in black on white or colored stock.

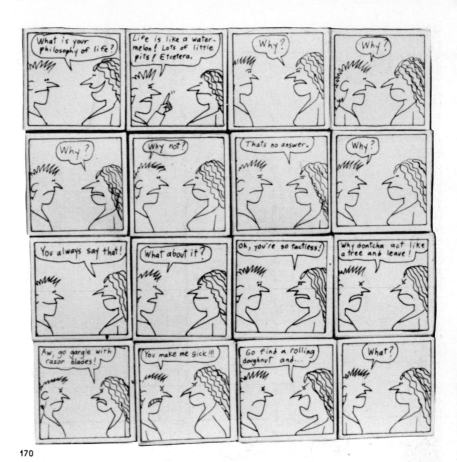

170

171

172

problem 6

Hotels and motels usually have a "Do Not Disturb" sign that can be placed by guests on the door. Create unusual, humorous, or decorative signs that can be sold on the retail level in stationery stores as a greeting card that can be used later as a doorknob sign. Signs will be sold with envelopes.

SOLUTION: There are several interesting and amusing solutions, one being the use of old prints to create unusual effects. These black-and-white prints are available from special "copyright-free" books. Photographs, old and new, can be most effective. Some design solutions call for a full-color, decorative design. Here are some of the solutions for this pleasant problem.

173

174

175

problem 7

Holiday gift cartons can be an expensive proposition for a manufacturer or distributor of products, especially in the gift, liquor, and cosmetic businesses. Find ways of dressing up the all-year-round carton to create a festive holiday look.

SOLUTION: The most logical solution is the "holiday sleeve."

The sleeve is a die-cut carton without a top or bottom. It can be fitted over the regular carton and removed after the holiday season. The sleeve is less expensive than a full-size carton or a decorative wrap. A variety of graphic effects can be created with clever die-cutting, embossing, textures, foils, and laminates. Even the less expensive types of sleeves create a festive holiday look.

176

with glass

problem 8

NO. 1. Some products, like nail polish, are usually sold with a remover. Since both items come in glass bottles, it would be an interesting project to develop a "twin bottle," consisting of two bottles joined together in some manner.

NO. 2. Some bottles contain liquids that have heavy sediments. Instead of the usual "shake before use" label, can you devise a better method?

SOLUTION 1. Two bottles with individual closures can be connected with a common "open end" closure. This is a tubular device with the proper threads to attach both bottles (with their own closure already placed over the bottles). The result is an "hourglass" design consisting of two identical or different bottles. Two jars or other containers can be connected in the same manner. The system enables the store to sell two items instead of one.

SOLUTION 2. An "upside-down" bottle would solve the sediment problem. A bottle rests on an oversized closure. The sediment descends toward the opening of the bottle. Once the bottle is turned over, the sediment will evenly mix with the liquid. This type of bottle or container would be most suitable for pharmaceutical products and paints that require mixing or shaking.

177

problem 9

Work up a design to recycle stock bottles and other glass bottles for a second use, preferably as a giftware or inexpensive novelty item.

SOLUTION 1. Convert bottles by screening or labeling to create an interesting and amusing bud vase.

SOLUTION 2. The same decoration method can be applied to design a new approach to greeting cards: "Greetings in a bottle." A paper flower with a wide stem can be inserted into the bottle with the message printed or written on the stem. Bottle can be sold with a recycled mailing carton.

178

with plastics

179

180

problem 10

Create inexpensive, reusable and practical packaging for children's confectionary, toiletry, and pharmaceutical products.

SOLUTION 1. There are some colorful, hollow eggs made of polyethylene. These eggs come in various sizes from 1-1/2"-4". They can be filled with candy, a small bottle, or a bag. The eggs can be decorated (screened) with faces or features of amusing people and animals. With a small weight in the bottom of the egg, the egg will rock back and forth like a "roly-poly."

181

182

SOLUTION 2. Another concept is a "bean bag" doll made of soft, fabric-type plastics. Parts of the body are filled with bath salts or pellets. Each part of the doll is a separate bag (that is, detachable) with a small opening to dispense the contents. After use, the parts can be refilled with real beans or peas to create a bean-bag doll. Scented powder can be used, and the doll becomes a pomander, ready for your clothes closet.

183

184

problem 11

Collectors of miniatures would welcome some type of box or case to display their collection.

SOLUTION 1. A plastic or wooden memorabilia box with adjustable shelves or partitions. The collector could arrange and display objects on the shelves. The box can hang on the wall or stand by itself. Items could be sold with the box in a heat-sealed, tray-type carton.

SOLUTION 2. A decorative, shadow-box display made of sheet plastics to show off individual items. The display can be used as the package itself, overwrapped and heat-sealed.

185

problem 12

Devise a "grow-and-shrink" package that would expand and collapse to hold a variety of "fun and games" merchandise for adults, as well as for children.

SOLUTION: Use the principle of the telescope. This package can be devised from extruded tubes (square or round) that slide into each other. An infinite variety of horizontal and vertical shapes can be created. The merchandise or components are inserted into the removable head.

186

187

with a variety of materials

188

Plants are sold in small or large pots or in planters. Some plants grow very fast and require some sort of stick to hold up the stem. Can you design a stick of some unusual quality and package it?

SOLUTION: A wooden figure with movable arms and legs to insert into the pot can be used outdoors and indoors. Since the figure can be bent into various positions, the height and width can be regulated. For larger plants, use two or more figures. It is packaged with seeds in a long narrow bag.

problem 14

What would be the best and the most entertaining package in which to sell medication for small children?

SOLUTION: A small, fuzzy fingerpuppet holds the bottle of medication. A small, colorful folding carton will hold the puppet.

189

190

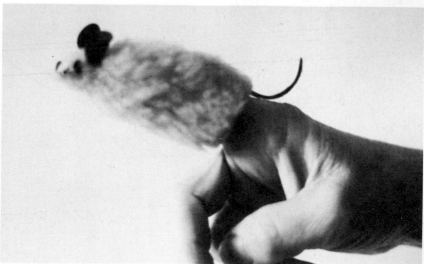
191

students' work

192 Christmas gift in shippable carton.

193 Game package based on a museum exhibit.

194 Gift item for the elegant chef.

195 Mail-order promotion for medical professionals.

196 Promotional package ties in rainbow theme on outside to rainbow product on inside.

192

193

194

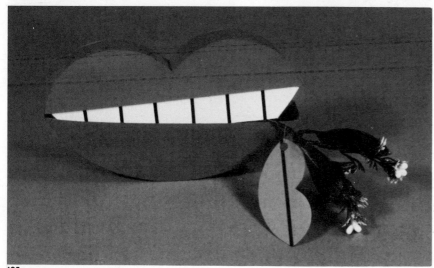

195

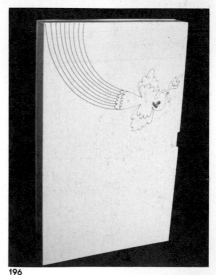

196

197 Package designs for all
aspects of a fast-food
service.

198 A half-dozen-egg crate
made out of E-flute corrugated
board.

199 Mailer box containing
vial of medicine
and promotional literature.

197

198

199

APPENDIXES

industry organizations

American Paper Institute
260 Madison Avenue
New York, NY 10016

American Management Association
135 West 50th Street
New York, NY 10020

American Society for Testing Materials
1916 Race Street
Philadelphia, PA 19103

Association of Independent Corrugated Converters
2530 Crawford Avenue
Evanston, IL 60201

Can Manufacturers Institute
1625 Massachusetts Avenue N.W.
Washington, DC 20036

Cosmetic, Toiletry and Fragrance Association, Inc.
1625 Eye Street, N.W.
Washington, DC 20006

Fiber Box Association, Inc.
224 South Michigan Avenue
Chicago, IL 60604

Glass Packaging Institute
1800 K Street, N.W.
Washington, DC 20006

The Fragrance Foundation
116 East 19th Street
New York, NY 10003

National Association of Recycling Industries
360 Madison Avenue
New York, NY 10017

National Canners Association
1133 20th Street, N.W.
Washington, DC 20036

National Center for Resource Recovery
1211 Connecticut Avenue
Washington, DC 20036

National Flexible Packaging Association
12025 Shaker Boulevard
Cleveland, OH 44120

National Paper Box Association
231 Kings Highway East
Haddonfield, NJ 08033

National Soft Drink Association
1101 16th Street, N.W.
Washington, DC 20036

National Solid Waste Management Association
1120 Connecticut Avenue, N.W.
Washington, DC 20036

Package Designers Council
P.O. Box 3753
New York, NY 10017

Packaging Education Foundation
1700 K Street N.W.
Washington, DC 20006

The Packaging Institute USA
342 Madison Avenue
New York, NY 10017

Packaging Machinery Manufacturers Institute
200 K Street, N.W.
Washington, DC 20006

The Paper Bag Institute, Inc.
41 East 42nd Street
New York, NY 10017

Paperboard Packaging Council
1800 K Street, N.W.
Washington, DC 20006

Society of the Plastic Industry
250 Park Avenue
New York, NY 10017

Point-of-Purchase Advertising Institute, Inc.
60 East 42nd Street
New York, NY 10017

useful periodicals

Food and Drug Packaging
777 Third Avenue
New York, NY 10017

The Glass Industry
777 Third Avenue
New York, NY 10017

Modern Packaging
205 East 42nd Street
New York, NY 10017

Package Engineering*
270 Paul Street
Denver, CO 80206

Package Printing
401 N. Broad Street
Philadelphia, PA 19108

Packaging Digest
410 North Michigan Avenue
Chicago, IL 60611

Paperboard Packaging
777 Third Avenue
New York, NY 10017

Product Marketing
124 East 40th Street
New York, NY 10016

*NOTE: MODERN PACKAGING AND PACKAGE
ENGINEERING merged in January 1980.

packaging education

Clemson University
Clemson, SC 29631

Essex Community College
Baltimore County, MD 21237

Indiana State University
Terre Haute, IN 47809

Joint Military Packaging Training Center
Aberdeen Proving Grounds, MD 21005

Michigan State University
East Lansing, MI 48824

New York University
School of Continuing Education
360 Lexington Avenue
New York, NY 10017

The Packaging Institute USA
342 Madison Avenue
New York, NY 10017

Fashion Institute of Technology
227 West 27th Street
New York, NY 10001

Parsons School of Design
66 Fifth Avenue
New York, NY 10011

Pratt Institute
DeKalb Avenue and Hall Street
Brooklyn, NY 11205

Rochester Institute of Technology
Rochester, NY 14623

Rutgers — The State University of New Jersey
Busch Campus
Piscataway, NJ 08854

Sinclair Community College
Dayton, OH 45402

University of Missouri-Rolla
Rolla, MO 65401

University of New Haven
West Haven, CT 06516

University of Wisconsin-Stout
Menomonie, WI 54751

employment opportunities

food products

REQUIREMENTS: Strong
typography, lettering, layout,
knowledge of printing
processes, plastic and metal
packaging.

EMPLOYMENT: About 85 percent
with top design organizations;
highly competitive.

beverages

REQUIREMENTS: Strong lettering
and mechanicals. Structural
design ability for
point-of-purchase materials for
the liquor and soft drink trade.

EMPLOYMENT: About 10 percent
with parent company and 90
percent divided between
design organizations and
suppliers. Knowledge of glass

packaging and model-making
skills helpful. Highly competitive.

houseware products

REQUIREMENTS: Basic packaging
skills, strong in printing processes
and film packaging (vacuum
forming, etc.), mechanicals,
typography, and lettering.

EMPLOYMENT: About 20 percent parent company, 65 percent design organizations, 15 percent freelance. Not too competitive.

health and beauty aids

REQUIREMENTS: Excellent taste, knowledge of current fashion trends. Top portfolio of original ideas. Strong in lettering, type, design, and layout. Knowledge of all printing methods and preparation of artwork. Familiarity with plastic, metal, and glass fabrications.

EMPLOYMENT: 35 percent parent company, 10 percent suppliers, 45 percent design house, about 10 percent freelance. Extremely competitive and selective field.

apparel

REQUIREMENTS: Good design ability with color, type, and renderings. Fashion illustration background helpful. Knowledge of all printing methods and preparation of art (mechanicals).

EMPLOYMENT: About 40 percent supplier, 40 percent design houses, and 20 percent freelance. Limited jobs.

industrial products

REQUIREMENTS: Strong understanding of technical data. Ability in product rendering and drafting. Knowledge of all package fabricating methods (plastic molding, glass, metal, etc.). Model-making ability.

EMPLOYMENT: About 55 percent with design organizations, 45 percent with suppliers of industrial packaging products. Competitive.

leisure time products

Includes toys, playthings, sporting goods, arts and crafts, giftware, etc.

REQUIREMENTS: Multiple talent and basic packaging skills. Good lettering, typography, layout; above all, design and illustration ability. Photography. Product design, model making. Knowledge of printing and plastic molding processes.

EMPLOYMENT: About 40 percent parent company, 45 percent design organizations, 15 percent suppliers and freelance. Not too competitive; good start for a talented graduate.

promotional packages

REQUIREMENTS: Strong background in advertising design skills. Ability in three-dimensional design. Printing and plastic-molding methods. Knowledge of photography and photo processing methods. Top creative ability.

EMPLOYMENT: About 80 percent design organizations, 10 percent supplier, 10 percent freelance. Fairly competitive.

point-of-purchase (POP)

REQUIREMENTS: Thorough knowledge of all media, printing processes, and molding in all plastic materials. Typography, lettering, and good mechanicals a MUST. Photography and rendering (comps) in all techniques. Model-making ability, especially working with cardboard and corrugated board, most helpful.

EMPLOYMENT: 60 percent manufacturer of POP products, 20 percent parent company, 20 percent design houses (special). Good opportunities for graduates, not too competitive.

glossary

aerosol An air-tight container with a dispensing valve and a pushbutton. The container is pressurized with a propellant gas which forces the contents from the container when the dispensing valve is opened. There are three types of propellants: fluorocarbons, hydrocarbons, and inert gasses.

ampoule A small glass or plastic container with a narrow stem or neck which is sealed after filling. Used for drugs, serums, and food ingredients. Opened by breaking the stem.

blister pack A transparent thermoformed shape (blister) attached to a card. The item is sealed between the shape and the card. There are several variations of this system.

blow molding A process for shaping plastic. Air is blown into a blob of molten plastic which is inside a mold. The plastic expands until it reaches the walls of the mold, to take the shape of the mold.

carboys Large bottle-like containers made of glass or plastics, usually encased in wooden outer crates. Used for shipping chemicals, water, and other liquid industrial products.

cartons Basically of two types: the collapsible folding and the rigid set-up paper box. Both types have several variations.

chipboard Recycled paperboard. The lowest-cost board, adaptable for special lining.

closures Closing or sealing devices for bottles and jars (glass or plastics). There are several types, including safety closures, dispensers, and applicators.

corrugated paperboard A construction of alternated layers of flat and fluted paperboard. There are A, B, C, and E flutes. Several layers can be combined for extra strength.

crimp To squeeze or press the ends of tubes or cans, by using a series of folds or corrugations.

drop test A mechanical procedure to test the safety of a package's content in shipping.

extrusion A process used in the manufacture of rods, pipes, tubes, and film, whereby a molten substance is forced through a shaped hole. When the substance hardens, it assumes the shape of the hole.

films Transparent or opaque flexible packaging materials, manufactured by extrusion. (See Chapter 5, "Flexible Packaging.")

fluorocarbons Chemical compounds used as aerosol propellants.

foils Metalic-coated packaging papers.

folding carton See: Cartons.

heatsealing The application of heat to close or to create bags or pouches from plastic films.

hydrocarbons Chemical compounds used as aerosol propellants.

injection molding Molding method whereby melted resin is forced into a mold by a plunger. Used only for thermoplastics.

labels A variety of die-cut self-adhesive (removable or permanent) applications used to decorate or identify packages or products.

molded pulp Paper pulp compressed in molds to create shapes and forms.

multiwall bag A flexible packaging form consisting of several piles of paper, plastics, or foils.

newsback Chipboard with one side white, the other gray. Used for inexpensive cartons, die-cut items, etc.

pails Large plastic containers, with or without lids.

paper bags These come in several varieties, with or without side gussets (pleats), used for groceries, shopping, inner containers, and other purposes.

paperboard Sometimes called Cardboard. Laminated layers of papers in sheets of 0.012" (12 points) or more.

plastics (See Chapter 4, "Plastics Primer.")

pouch A flexible container made from films, foils, or paper.

recycling The reuse of previously used materials.

resins Raw material for plastics.

resource recovery The use of materials that otherwise go to waste.

shrink film A low-cost wrap, sealed with heat, for units or individual products.

skin packaging A method by which a thin plastic film is drawn over the product mounted on a card or sheet.

thermoform The shaping of heated thermoplastic sheets or films through forced contact with the mold.

tubes Extruded plastic or glass products used for packaging.

tubs Large plastic industrial containers.

tumblers Glass or plastic containers with metal or plastic lids.

universal product code (UPC) A code printed on packages that provides information on the product for inventory control and retail pricing.

vial A small glass container for medical or pharmaceutical products.

welding A method involving heating or spinning the edges of two plastic components and then joining them together.

bibliography

Environmental Protection Agency, Office of Solid Waste, Recovery Division, Washington, D.C.

Griffin, Roger C., Jr. and Stanley Sacharow, PRINCIPLES OF PACKAGE DEVELOPMENT. Westport, Ct.: The Avi Publishing Co., 1972.

MODERN PACKAGING, 1978, 1979, New York.

MODERN PACKAGING ENCYCLOPEDIA, New York.

"Municipal Solid Waste: Its Volume, Composition and Value." Washington, D.C.: National Center for Resource Recovery, Inc., 1977.

PACKAGE ENGINEERING, 1978, 1979, Chicago, IL.

PACKAGING DIGEST, 1976, Chicago, IL.

PACKAGING INSTITUTE USA, 1979, New York.

Packaging Systems Corporation, THERMOFORMING. Orangeburg, NY, 1978.

"Refuse-Energy Systems with Resource Recovery as Alternatives to Landfill." New York: White, Weld & Co., Inc., 1977.

Sacharow, Stanley, and Roger C. Griffin, Jr., BASIC GUIDE TO PLASTICS IN PACKAGING. Boston: Cahners Books, 1973.

Stanley, Thomas Blaine, THE TECHNIQUE OF ADVERTISING PRODUCTION. New York: Prentice-Hall, Inc., 1954.

"Third Report to Congress, Resource Recovery and Waste Reduction." Washington, D.C.: U.S. Environmental Protection Agency, 1975.

INDEX